To Be, and Not To Be

The Rise of Misplaced Power and What It May Foreshadow

A Novel

**VOLUME 1
(What Was)**

Robert M. Lebovitz

© Reg. 2014
Copyright 2016, 2018
Robert M. Lebovitz, Dallas, Texas

This novel is a work of fiction. Names, characters, characterizations, places, and events are products of the author's imagination or fictionally used. Any resemblance of the work's characters or locale to actual persons, living or dead, to actual places as they are or as they were, is entirely coincidental.

All rights to this work are reserved. No part of this work may be reproduced, stored in or introduced into a retrieval system, uploaded or transmitted in any form or by any means (such as but not limited to electronic, mechanical, photocopying, recording) without prior written permission from the author and copyright owner.

"What's Going On" and "Cage of Freedom," from *Metropolis* soundtrack album, used by permission of Giorgio Moroder/Soundtrack Music Associates.

Image of "The Wedding Dance" by Pieter Bruegel, the Elder (Detroit Institute of Arts, USA / City of Detroit Purchase) used by license from Bridgeman Images.

No portion of the cover art or any interior monochrome photographic images may be used in whole or in part without prior written permission from the artist and copyright owner, Robert M. Lebovitz (Eye2Eye Gallery, Dallas, Texas).

ISBN 978-1-7325045-0-9

To Be, and Not To Be

*The Rise of Misplaced Power
and What It May Foreshadow*

**VOLUME 1
(What Was)**

Lebovitz 1980

TABLE OF CONTENTS

PREFACE iii
ACKNOWLEDGMENTS ix

VOLUME 1
(What Was)

PART I

1. *Casablanca, Réchauffé* — 3
2. Anticipations of Sundae Nite — 17
3. Payback — 47
4. Who is Barnard? — 91
5. Sundae Nite — 133

PART II

6. Movie and a Visit — 161
7. Barnard's VieGie — 183
8. Considerations of Elder Eden — 199
9. Current Affairs — 239

VOLUME 2
(What Is)

PART III

10. Hutch's Conceit — 265
11. The Notice — 287
12. Barnard's Talk — 303
13. Phil's Insight — 327
14. Quantum Superposition — 347

PART IV

15. Bus Ride — 371
16. Taking Leave — 403
17. Connections Lost and Found — 431
18. Caol Ila 18 — 461
19. "Heeeeer's ..." — 487

VOLUME 3
(What Will Be)

PART V

20.	Metamedics	515
21.	A Frank Visit	539
22.	Ty's Remonstration	557
23.	Condi Redux	593
24.	Phil Misspeaks	603
25.	Weill Visits	635
26.	Venice in the Sun	645
27.	Hutch's Reproof	679

PART VI

28.	Soup, and Soup Again	691
29.	One Visit, in Two Scenes	707
30.	... and Not To Be	735
31.	Local News	765
32.	Barnard's Truths – Weill's Wiles	779
33.	To Be	811

AFTERWORD 825

PREFACE

A notable aspect of our milieu is the mutually reinforcing combination of computerization, media, and the World Wide Web. Information has never been so easily obtained and so universally accessible. It would be natural to think that we are better informed. Unfortunately, the scientific methodology is obscure and deeply buried, news and commentary are corporate and agenda driven, and the public Internet is overwhelmingly mercantile. While flooded with data, we are in many ways, if not less informed, certainly often studiously misinformed. Few acknowledge, however, that facts are labile and incontrovertible truths rare. Most believe that, in today's world, they are even better able to "know" and thereby guide themselves accordingly. Yet, to naively assert one is reliably informed by the plethora of available data and commentary is dangerous. It opens the door to manipulation and control. One must remain critical of "facts," which are often the servant of duplicity. *To Be, and Not To Be*, a fiction that has become dishearteningly current, attempts to explore the technologically altered landscape in which that task is paradoxically more difficult.

The protagonist's first level of difficulty is that, being elderly, the new world is far different from his old. He is obliged to adjust and to adapt. The second, again because of his seniority, is more sinister. There has occurred an unprecedented domestic financial crisis, one brought about by the collapse of our international stature along with our dollar. Being in the most senior segment of the population sets him apart. Once productive and contributory, his diminished utility makes of him, like virtually all his peers, a liability with assets. As its dramatic center, this novel explores the consequences of being senior and retired in a country that is unable to pay for entitlements once blithely enacted and, further, has come to view the vast store of funds, which that segment of the population has put away for retirement, as a valuable short term resource, one that should be "... put to better use." *To Be, and Not To Be* is, in short, a story of using technology, coupled with necessary and sufficient administrative reasons, to make victims of those who had previously been beneficiaries.

Barnard Cordner – former physics student, naval officer, then Professor of Economics, and now retired – must contend with governmental machinations that are life changing. He grasps the government's stated economic justifications for the relocation of seniors and the expropriation of their retirement funds. He just does not believe them. By temperament and training a rationalist, he refuses to accept that authoritarian pronouncements should outweigh verifiable reason. Unable to take for granted what he believes he knows or how he knows it, he is presented with a host of uncertainties, which he would prefer to resolve but cannot.

Uncertainty greatly complicates our lives. Therefore, some see its minimization via proscriptions and extrinsic control as a positive step forward. Overbearing guidance and technological manipulations thus can be presented as social benefits rather than the impositions they truly are. As the protagonist muses in the novel's penultimate, revelatory chapter, "No need for reasons. No need to ask Why when everything's managed." Being truly free carries the burden of needing to deal with uncertainty, to acknowledge and survive with it. *Either/Or* is the title of a dense philosophical tract dealing with personal uncertainty. Hamlet's pithy soliloquy, "To be, or not to be," has a similar focus but is embedded in drama and thus far more accessible. However, both works at least hint that an answer must exist, that one choice must be right and the other wrong. Naive intuition leads to a similar conclusion, namely, that a fact, a statement, or an observation is either true or it is not. But what if such separability does not exist? What if reason alone is not adequate to resolve "... the question?"

The theme of this narrative is that the quandary in which the protagonist finds himself, as indeed we may also, is more akin to the superposition concept formalized in quantum mechanics: His personal Is and Is Not coexist until resolved by some manner of test. It is not a matter of one or the other – Hamlet's exclusive "or" – but rather one of essential commingling, an ambiguous "and" that awaits resolution. This is but one of the paradoxes that confound fully understanding the physical realm. Experimentally demonstrable and thus hard to deny despite that it defies personal experience, it is an apt counterpart to the dilemma here dramatized. What techno-social forces are using to their unique advantage in this novel and the reason for the protagonist's angst is that, until tested, Is and Is Not can be made inseparable as well as indeterminate. This makes a priori assignment of these once disparate logical alternatives irrelevant, which can be worse than being unresolved. Sensing inherent irresolvability yields more anxiety, in other words, than does simple uncertainty, since it takes away hope. After-the-fact, a posteriori determination being all that can be known and, therefore, all that matters, leaves one adrift, at the mercy of those who can confidently mask their own ignorance with action and professed certitude. Accustomed to the anticipations of reason, the protagonist – a former naval officer – is here rudderless, yet forced to "... contend with a turbulent sea and an uncertain port."

A privately printed version of this novel was circulated in 2013; a hardback trade edition appeared in 2016. Subsequently, events, i.e., reality, have overtaken a number of its projections and fictional anticipations. The harsh, far right, self-righteous, and single-issue mood that has since emerged makes it imperative to consider: Where will we go from here?

To some extent, this updated version of *To Be, and Not To Be* is an attempt to do precisely that. In addition to revising title and text for style, recent facts are incorporated in this three volume paperback edition, primarily in Chapter 12, to ensure topical accuracy. Still, this remains a work of fiction, a fantasy of what may come to be. Current trends are clear; parallels with the past are unmistakable. It is not hard to imagine that, a decade or two from now, we could be uncomfortably close to or enmeshed in circumstances comparable to those presented here. Economists are already voicing their fears of a coming severe debt crisis. Our long-term future will depend upon how we, as a society, deal with any such severe economic dislocation, on whether we do actually experience a historic fiscal emergency, one that could compromise participatory governance – democracy – as we have come to know it. But similar paths, even similar societal forces, do not necessitate similar ends. There is always hope.

That many aspects of this novel are now recognizable in the media suggests veracity. Consistent with its dramatic intent, however, this may be illusory and not survive critical test. In any event, I apologize in advance to those who may find it offensive. The latter is to be preferred over the tragedy of having its dark potentialities realized. To the degree that this has already occurred, the subtitle is superfluous.

The phrase "misplaced power" comes, of course, from President Dwight D. Eisenhower's Farewell Address to the Nation (January 17, 1961). A portion of his speech is worth quoting at length:

> "This conjunction of an immense military establishment and a large arms industry is new in the American experience. The total influence – economic, political, even spiritual – is felt in every city, every Statehouse, every office of the Federal government. We recognize the imperative need for this development. Yet we must not fail to comprehend its grave

implications. Our toil, resources and livelihood are all involved; so is the very structure of our society.

"In the councils of government, we must guard against the acquisition of unwarranted influence, whether sought or unsought, by the military-industrial complex. The *potential for the disastrous rise of misplaced power exists* [emphasis added] and will persist.

"We must never let the weight of this combination endanger our liberties or democratic processes. We should take nothing for granted. Only an alert and knowledgeable citizenry can compel the proper meshing of the huge industrial and military machinery of defense with our peaceful methods and goals, so that security and liberty may prosper together."

To apply President Eisenhower's observation to the present, we need only put in place of "industrial," "military," and "machinery of defense" the words mercantile, informational, and Internet. Eisenhower's warning, thus suitably transformed, is applicable to the richly interconnected world in which we are embedded. Eisenhower then went on to say:

"Yet, in holding scientific research and discovery in respect, as we should, we must also be alert to the equal and opposite danger that public policy could itself become the captive of a scientific-technological elite."

We are in the grip of a powerful info-mercantile Internet complex. Its reach is pervasive, its concentrations of power are unprecedented and its manipulations, by design, only dimly perceived. There are, indeed, grave implications, of which many of our scientific-technological elite have offered explicit warnings. Equally vocal have been those who impugn these dire predictions. Both poles of this futuristic commentary need to be judged in the context of economic self-interest. Too many hold that their mercantile success is ample evidence of the validity of their projections, projections which often are more self-serving than enlightening.

However, we must never yield to the canard that advancements of knowledge and technology are inherently evil, despite what those fearful of displacing ignorance, superstition, and myth may claim. The evil resides in how that knowledge is used. The dark side of technology emerges from tendentious rationalizations for its misuse, from how circumstance can contort its application to serve narrow and special interests. This is what we need fear.

We have thus far been fortunate. No such dire pressure – be it economic, societal, or political, be it attributable to trend, to the inevitable social cycles, or to the Hegelian formalism of thesis-antithesis-synthesis – has reached a critical stage here. There can be no doubt that such stresses are building. Whether domestic or foreign, whether truly within our ability to control or not, the consequences of facilitated overreaction could be seismic and irreversible. Therefore, where the info-mercantile complex could lead us, where it *is* leading us, should be of concern. The dark potentials, of loss of self as well as loss of control over its definition, need to be explored. If plausible fiction serves to do so, then it should be used.

Dallas, 2018

*   *   *   *   *

Why, then, 'tis none to you; for there is nothing
either good or bad, but thinking makes it so.
Hamlet, William Shakespeare

*   *   *   *   *

"Men's courses will foreshadow certain ends, to which, if persevered in, they must lead," said Scrooge. "But if the courses be departed from, the ends will change. Say it is thus with what you show me."

The Spirit was immovable as ever....

"Spirit!" he cried, tight clutching at its robe, "hear me. I am not the man I was. I will not be the man I must have been but for this intercourse. Why show me this, if I am past all hope?"
A Christmas Carol, Charles Dickens

* * * * *

... [my generation] learned that one can be right and yet be beaten, that force can vanquish spirit, that there are times when courage is not its own recompense.
>Albert Camus, in speaking of the Spanish Civil War of 1936-9
>(attributed to Preface to *L'Espagne Libre*, 1945)

* * * * *

Is there a mask behind the face?
Is it a sleeveless ace?
Can a smile conceal a sneer?
How do we see clear?
Answers can change the question line
Every time
And now the truth in confidence
Tells a lie ...
What's going on? I wanna know

Cage of freedom, that's our prison
We fabricated this world on our own
World on our own, world on our own
>*Metropolis*, Music - Giorgio Moroder, Lyrics - Pete Bellotte

* * * * *

ACKNOWLEDGMENTS

No comprehensible fiction comes forth without ties to fact. Classic and acclaimed prose, while generally separable from real events, links to the observations, representations, and analyses that have found their way into the body of useful knowledge. I am, therefore, deeply indebted to many giants of literature by virtue of having had the privilege of reading their works. It is a never ending task.

I owe thanks to many for reviewing the first (2013) version of this book. In particular, the critical comments from Prof. Paul Geisel and Dr. James Wagner were detailed, insightful, and, therefore, most helpful. Any errors (of past, present, or future fact) that remain or were inserted thereafter are my own. Thanks, also, to Ms. Susan Schewe who, through her thoughtful initial editing, helped me to clarify the narrative. Above all, I want to express my most profound thanks to my wife, Joyce Sanders – my partner in love and life – for her patient perusals, for her many thoughtful suggestions as I wrote and continue to write.

PART I

Lebovitz 2009

CHAPTER 1

CASABLANCA, RÉCHAUFFÉ

There was no need for Rick Blaine to have died that way. I know. That's Hollywood. It's what they do. It was entertaining, right, but different somehow. *Casablanca* just didn't feel the same, not on that recent viewing at any rate. I really wanted a different ending.

I've a long list of favorite films, most from years ago. While their entireties have faded, I do recall highlights, iconic scenes. When sleep is elusive, I'll replay these to mask difficult thoughts. Films recently or repeatedly seen, via cable, DVDs, streaming downloads, or on the net, can be more easily revived. I might start with an exceptionally good opening hook and hopscotch through an entirety. The disorderly pieces of a whole, its barely scripted incidentals, usually suffice. Recollection, you see, fills in the gaps, takes the place of cinematic priors and thereafters.

But I prefer the old, the well known, over the new, the recently judged. Motes seem to survive even though immersed in banality. "You know how to whistle, don't you, Steve? You just put your lips together and … blow." I'll never forget, nor fail to mimic on the proper occasion, that promise to "... be bahk." I'll forever smile at the thought of the Fat Man's incongruously brisk stride in *The Maltese Falcon*. In its concluding scenes his paunchy disappointment at the reveal of the false bird is undeniable, his maniacal, slashing descent from expectation into despair is explicable, his final transfiguration to reinvigorated optimist plausible. Of an entirely different genre is the lilting "Daisy, Daisy, give me your ..." with HAL's slowing, distorted bass playing against the rhythmic hiss of a human's constrained breathing. Recently, last year in fact, we've heard the hoarsely whispered "Do it! Just damn do it!" of an aged, born-again Bourne nee Damon. Or go back, far back, to the mask ripped away with neither shriek nor gasp. The device of silent surprise worked well then; it's too simple a fright for most now.

Awakened emotive connections underlie the power of narrative based upon moving images and coordinated sound. From a properly constructed sequence we experience excitement or fright, joy or envy, elation or sadness. "Hello, you old Savings and Loan!" in *It's a Wonderful Life* conjures up sympathetic relief during any Christmas week; the tinkling tiny bell in its finale is sufficient to elicit misty eyes. In heroic epics we partake of the experience of overcoming or failing. After any good horror film we're prompted to attribute dark significance to footsteps behind. No matter what the style of production, the point of a film is to convey a meaningful, even if understood to be false, reality.

I know. It's regressive to persist in using the word "film" in this world of bits and bytes. Old habit, I suppose.

For a cinematographer, being "e"ffective is primary. Illusion is the key to unlocking connection. The aim, however, should be to be "a"ffective, to elicit emotional connection beyond mere faithful representation, beyond meaning and conventional reality, beyond words and formal truths. Via images on a screen or words on a page or space-occupying art, the skillful do move us beyond facile recall and raw apprehension of facts. Those with lesser skills depend upon explication. They have only brief relevance or fail totally despite, or possibly precisely because of detailed annotation. The best succeed without it. The intellectual insights their works evoke follow rather than lead, are incidental rather than revelatory, responses to rather than causes of. Significance, I mean to say, should be a consequence of the emotions aroused, not their cause. Some artists have an instinct for attaining it repeatedly. Others stumble onto good fortune, create an isolated success that their subsequent work reveals would've been wiser to have let remain so.

Right?

But that's another matter, the craft of professionals. It's my own altered affective response that I'm struggling to understand. Some of my favorite films feel different, their emotional impacts seem altered. I find this troubling since, as much as these could be signs of personal growth, so could they be symptoms of regression. I can accept the fading of remembrance. I'm old enough. Still, I'm of the opinion that affect, if retrievable, if evocable at all, should not change. Yet, some familiar plots have come to lack the expected, previously satisfactory thematic arcs.

A perfect example is the film *Casablanca* or, I should say, my reactions to it. As many times as I've watched that piece of cinematic escapism, that mildly propagandistic story of clear-cut heroes and villains, it always had elicited feelings of adventures well-met and yet to come. This last time, just last week, it didn't. The ending struck me as cold, didactic, devoid of promise. It was appropriate to the plot and internally consistent but, at the same time, teasingly unsettling. It left an itch of unsatisfied connection. My reaction to *Casablanca's* resolution was, and remains, as much disturbing as unexpected.

The consummation of *Casablanca* is, of course, that melodramatic quintet at the airport. First Rick Blaine and Captain Renault, then Ilsa and Victor Laszlo arrive, each anxiously anticipating the departure of the Lisbon plane. Major Strasser is en route, thanks to Renault's casual duplicity, only they don't know that. Tight shots of the iconic German officer in an open car are cut in to make evident his anxious determination. On the tarmac in front of the hanger, Renault does Rick's forceful bidding. Ilsa resists. She can't accept that he is sending her off with Laszlo.

"But, Richard, no, I, I –," she says.

Yes, they'll always have Paris. The close-up of her upturned face, the moist eyes and faintly parted, pouty lips, worked its magic on me. I was unabashedly moved. I blinked back tears, as always. I took a deep breath, as always. As Ilsa and Victor blend into the mist, I felt glad for them.

Were *Casablanca* to be made today, some tedious, over-hyped hack would no doubt insert a song, probably entitled "Last Plane to Lisbon," using ascending chords and a terminal minor, thereby asserting the device of plot emphasis but primarily striving for secondary residuals.

Hah!

Anyway, jumping out of his car at the airport, Strasser confronts the pliant Renault and the resolute Rick, who are still gazing into the milky mist. When informed that "Victor Laszlo is on that plane," he marches to a phone box with briskly conveyed, evil intent. He's sharply defined, monadic. He grabs the cumbersome, black handset.

"Put that phone down!" Rick growls.

"Get me the radio tower," Strasser demands, his expression a rigid mask of rage. His hand explores his coat with obvious intent.

There's another unheeded command from Rick. "Put it down!" he says, with his so distinctive intonation.

Rick's pistol is at the ready. Strasser's Luger is a hastily retrieved blur, but unmistakable. That quintessential Nazi sidearm was a staple of the Us versus Them films of that era. They both fire. Strasser's trigger pull is quicker, his aim inexplicably surer. We don't see Rick take the bullet, but we hear it. We hear also his grunt and the clatter of his gun on the tarmac.

I've seen Bogart die in other movies and so could imagine his toothy grimace. Rick inexplicably hesitated. He considered options. Strasser did neither. From behind, his slight jerk then, for a frontal frame or two, the brief shock reflected by momentarily closed eyes suggest a superficial wound. Not so for poor Rick. Strasser's shot, even while barely aimed, is fatal.

I appreciate the entertainment value of secondhand reification, the re-experienced impact of that which is itself a fiction. It's what keeps classics in circulation. Being taken back to a known moment, whether it be real or fictional, can be rejuvenating, reassuring, well worth forsaking novelty. Eidetic imagery can take us farther in. Still, even that only partially travels the road to full veracity. Remembering, viewing a media image of, or internally recreating a representation of place or thing or event, no matter to what extreme of faithful detail, is notably different from experiencing it again. That requires a coordinated replay of emotional context. It's based upon establishing a complex internal state that may never have existed otherwise and, confounding many attempts to recreate it, is probably different in each of those impressed by otherwise identical externals. We aren't, after all, simply sensor-driven machines. No barrage of visual and auditory input will recreate reality unless it awakens latent, native, and, most often, individualistic emotions.

Psychologists and neuroscientists have suggested that olfaction, a most primitive sense, is powerfully endowed in this way. Even in the highly developed sapient brain, the sense of smell has retained its archaic anatomical connection to the cerebral circuitry controlling drive and the urges that we experience as emotion. Odors can bind to events, to specific places and times, thereafter be sufficient to reawaken their emotional contexts. I'm not unique, I'm sure, in having experienced this many times. The spicy scent of a hand-thrown pizza, for example, can transport me back decades, a half-century or more. Sometimes this is to a summer evening, standing alone, watching an expanding circle of dough being tossed. Or I may be momentarily on the bench backseat of a car by the reservoir, with the open, flat box still warm across our thighs.

Perfumers are adept at using the nose. The olfactory sense could someday provide the key to solidifying the truth of an emergent world.

Right?

Films and videos lack that, obviously. Surrogate representations via current media rely upon cleverly crafted audio and images. While the latter may be primary, a well designed sound track – with rich instrumentation, with clever chord progression and modulation of pace – can pull us further in. But it still doesn't guarantee we'll be full participants. To suspend the awareness of artifice, to truly feel a moment as fact not as recreation, there must be more. All of our senses, of course, contribute to the creation of our private realities. An odor would serve, as I've said, or perhaps making use of an articulated chair. If its motion is coordinated with the images, it only has to shift in small increments to make a person as dizzy, to give the same white knuckles and nausea as from a real midway ride. That's feasible right now and much more practical for home use than is olfaction.

We should never underestimate how much people will spend to reify their fantasies or how much time they will devote to doing so. There's a lot of money made from servicing that desire. Technology has merely added the means, the additional avenues, and will continue to do so. But, in addition to the feasibility and expense of fooling the sensorium, there is the role of memory. That's the point I want to make.

A fully endowed recreation, be it of fact or fiction, is an intricate tapestry woven from diverse strands, with many of its threads capable of providing access to the whole. For me, the cold pressure of metal against skin, the sound of an engine, or the motion of an unstable pier each can suffice to reestablish the richly real, multidimensional setting of a specific past moment and, thereby, bring it forward to the present. A light touch on the cheek, the contours of a body part or its surrogate, a musical phrase, the taste and mouth feel of softening ice cream can each go far beyond being simple sensory events. They reawaken prior realities.

For the illusionist, the trick, then, is to tug on such accessible links and thereby pull out affective arousal. Playwrights and directors do it repeatedly. Musicians and dancers, novelists and poets, lovers, masseurs/seuses and whores, strive for it. Artists, whether classical or contemporary, representational or abstract, have that aim, if you attend to their memoirs. Designers and chefs offer that they can achieve it. With focused research, all of the senses could someday be artfully coordinated to that end. At present, as I said, what we see, what we hear, that couplet of sensory modalities that are fully engineered and readily linked to an autobiographical past, must suffice.

I should clarify that by media I primarily mean the Internet. Certainly, its news, entertainments, and enlightenments are centrally managed. This has been a profound but necessary adjustment and one kept out of sight as much as possible. However, careful management has provided advantages that outweigh any understandable distrust of uniformity, lack of privacy, and control over content. From my perspective, it does offer much – yes, such as the prompt availability of movies such as *Casablanca* – that, usually with Sallie for company, are pleasant enough. These are sufficient to warrant ignoring the Internet's duplicity and intrusiveness. In time it no doubt will evolve beyond representing reality to defining it, to decoupling it from any originating external. That's something to consider elsewhere. Here I'll just say that I'm not always charmed by its busy images or by the salubrious tones of its interface. Its linkages are rarely mine. Also, this dystopic world's events are heavy enough without being drenched in factitious detail.

Right?

I'm forced to admit, however, that mine may be a generational judgment, one not shared across all age groups. The Internet is a valid tool, a means. Imparting a consistent reality – recreating and conveying it – is its primary task. While I understand this, it's not my profession. I'm not a media specialist, as is probably obvious. I am, I was, I should say, an economist. I have little standing, therefore, to be expounding at length upon that which creative people grasp instinctively. We all harbor a Madeleine. The Many, those busily engaged with the Now, survive quite well enough without noticing. I generally do. I notice, I mean to say. As sophomoric as may be my elaboration, it has personal relevance. I've cause to be perturbed when I find that affective memories are not reliable, aren't always familiar.

That recent, last viewing of *Casablanca*, for example, raised exactly that issue. In other situations, as well, I have found certain memories more sui generis than déjà vu. This may be due to the unsettlement of age, or the result, again, at my age, of having excessive time to dwell on it. Either, unfortunately, would suffice. The even less welcome possibility would be that my ability to recall is impared. From conversations with other retirees here at Sunset Vistas, I suspect this isn't so infrequently noticed as it is imprudent to remark upon. In any event, the test, the validation of reality should not be via memory. The reverse should be the rule.

Well, I suppose I should refocus on my main point.

Movies become classics because of the stable truths they evoke. Whether intent is there or not is irrelevant, except in argumentative judgment of the worth of the filmmaker apart from that of his creation. *Casablanca* has always exemplified for me how a stream of emotive vignettes, if deftly portrayed in sympathetic contexts, can elevate even suspect narrative. Even long before I retired, when having a dozen or so channels seemed a boon, if some nonattributable inner turmoil confounded my attempts drift off, then a familiar story, one with a plot so porous that I could enter at any point without substantial loss, was the ideal soporific. Later, the need might've arisen while awake with Diane quiet and utterly blameless beside me. She could sleep easily before the real pain came; I never was an easy sleeper. *Casablanca,* as unencumbered by weighty confusions as is its protagonist, Rick, was easy to engage. It was always sufficient to nudge me away from overindulgence in the ongoing minutiae of politics, money, or academe. Its familiar story line dispersed the miasma of the vexatious, relentlessly churning, fanciful images of what had been, what is, and what will be. It served to displace, if only briefly, the fears of Diane's unalterable course.

I recall years ago stumbling upon a network Bogart festival one weekend, while lying about and profoundly lethargic – from strep or mono, I can't remember which – and bored with overly annotated sports. *The Big Sleep*, the sound-studio gunplay over, was nearing its conclusion. I had little sense of loss over the unseen preliminaries. Bacall with Bogart was enough. During the interval they stated that Bogey had enjoyed sailing. From this I imagined a connection between us. I vowed to learn where his craft had been berthed, whether it had been, like ours, at Del Rey. Then came *High Sierra* and, finally, *Casablanca*! It was as affective as it was contrived. Concern with the underlying patriotic intent had long since faded. The group sing of "La Marseillaise," for example, to diminish the hands-on-hips Fascists, affirmed the back story. Nor were lapses of attention detrimental. As the perception of movement is created by a rapid succession of stills, intermediates supplied by memory plus an instinct for continuity convoyed the film's notable dramatic moments forward.

1 -- CASABLANCA, RÉCHAUFFÉ

The visual quality of streamed movies is much better than the old tapes and discs. I'll admit that. And you can get them on a whim, can follow any spontaneous urge. Yes, the government monitors the streaming downloads, as it does 'net services in general. It has good and sufficient reasons, its own interpretation of exigent circumstances. Right? But that, also, is another topic. I'm not going to rehash that here. It suffices to say that technological advancement provides a plethora of benefits – especially if one emphasizes the trivial, I might add. In any event, when I last had that urge to escape, when I wanted to watch something devoid of current themes and exhortations, I scheduled *Casablanca* for immediate view.

For the most part it worked. However, after the closing credits, as I stared at the menu on the big screen, I was emotionally disoriented. The movie was cognitively familiar but did not *feel* the same. It wasn't confusion or a sense of error. Wrong was not the word, even though it left me with the powerful impression that a different ending would've improved it. That had never been the case before. It always had been satisfactory as it was.

It was a treat to again experience the novelty of black and white, the smooth, masculine Humphrey, and the gorgeous, compelling Ingrid. She doesn't appear nearly enough but when she does, oh my: the diffuse lighting, in general; that puffy delight of her lower lip, in specific; the hint of a tear. I used PAUSE several times to fix her on screen for a long look. Years ago I had come across a *For Whom the Bell Tolls* sound track long play, an actual twelve-inch vinyl original with her face on the album cover. That's it. Just her teary, far from Nordic-toned face – no title or annotations of any kind. I had no interest in the music, kept it pristine in a plastic sleeve for years.

Actually, there are numerous bits in *Casablanca* that were worth revisiting. Take, for example, the few earnest lines spoken by Annina, the young woman so anxious to escape from Eastern Europe. Her reply to Rick's dismissive, "Go back to Bulgaria," is, "Oh, but if you knew what it means to us to leave Europe, to get to America!" She overrides his sarcasm with the perfect blend of desperation and determination. I relished the mellifluous Claude, the subtle ease with which the demigrandee is shocked by the obvious. I looked into Lorre's anuran eyes. I smiled at Sidney Greenstreet's snorts and tummy chuckles, at Kinskey's droll take on the archetypical Sacha, at Sakall's channeling of Old World attitude via Carl's artful shrugs.

The current evening video offerings aren't nearly as pleasurable for me. Inherently so, probably, because I'm neither a productive consumer nor a targeted viewer. Crappy tribute specials annoy me, they don't entertain. Decrepit, retro-dressed *älters* feebly channeling instances of their former personae? Bah. Totally tedious. The LOL Show – a collection of barely rewarmed comical pap – is another example. Rereruns – laughably oxymoronic, if not overtly moronic – of *SIC-Moscow* or *Istanbul After Dark* are okay, even with their multilingual patter and recycled plots. Unfortunately, their device of putting the viewer in the midst of the plot, or even being ahead of the characters, is now so overdone as to be self-defeating. As to the full length features, most are mainly noise and special effects, overly loud, hyper-energized treatments for adolescent minds in two decade old bodies, and so full of messages and misdirects that, having lived a long life, the transparent intention to manipulate is painfully apparent. I so prefer aged classics over banal, sense blasting, short attention span crap that –

Shit. Enough of that. That's old news.

What I'm trying to explain, is that the other week, that last time I actually viewed the film, *Casablanca* struck me as different. That's my point. That's the core of my unease. It was familiar, but it didn't provide the ending I wanted, not the one I felt Bogey's Rick Blaine deserved. Nevertheless, it was what it was.

Actually, neither was it what I wanted for Diane. That, too, was what it was. She, too, deserved better. After the pain dug into her and took over her, I couldn't watch TV at full volume if I passed a night on the couch. She needed every bit of absence she could manage. Muting was too extreme, since I had to be able to glean at least a hint of the emotive tonality of the dialogue. The words themselves were for that other aspect of the brain, the objective part. I suppose I could've used earbuds. Except, besides being irritating, I might then not have heard Diane call out. With her fitful on the other side of the wall, which I had double paneled specifically to muffle sound under the pretext of supporting the frequently moved framed art, I came to rely upon the supplementary streaming text.

After she died, I was released from caring tasks and sad reminders. Superficial escapism was less compelling. I could fill a long night with a good read or trying to work on my long-planned economic text project. I became accustomed to being alone. Moving to Sunset Vistas was a major change. There meeting someone as sweet and uncomplicated as Sallie was my unanticipated good fortune. She coaxed me back from emotional isolation. While we lack a shared history, which deficiency may be outweighed by an uncertain future, we do have the present. There are no grand gestures binding us; what we have is companionship seasoned with mutual regard. Occasionally, even that degree of closeness is not what I seek. That's how I am. It's hard to totally excise singularity, or should I say, reinforced solipsism. It's how we all start out, after all. It's only from experience that, having come to think of ourselves as distinct from all else and having come to appreciate the boundary separating the two, we accept that there is validity on either side of it.

In any event, Sallie and I spend many evenings and nights together. I shouldn't dissemble. It's not lust. We don't do sex, if you're curious about the elderly. We're content with double entendres, fake tussles, and real cuddling. Neither is it companionship in any exclusive sense, since Sallie has May, and I, as I've implied, have grown accustomed to empty rooms. Enjoying something and missing its absence are not necessarily conjoined. Right? I've adapted. We both have. The pleasures we share are as ordinary and fundamental as sex: meals prepared in my adequate kitchen; the occasional outing; sharing adjacent warmth in bed, only incidentally observed; word play; watching a show or film. Our relationship is not virginal. We see each other, in the shower or when getting dressed in the morning if she's chosen to stay. We make jokes. We grab at each other or mingle our legs at night. We rest. We try to sleep. We deal with the nocturnal necessities. We don't conjoin.

To this day I find it difficult to state aloud that I sleep with her. That's the operative term for fucking, you see. Carnal knowledge is from another era, whereas 'sexual intercourse' and 'coitus' are clinical terms that eschew emotion. "Had sex with," is perfectly acceptable to say. It's more accurate than stating, "I slept with," yet doesn't sound causally crude nor crudely casual. It's much the same sort of replacement, I suppose, as is calling the penis a weenie. Personal realities can be difficult to express.

Again I'm digressing.

Sorry.

I should focus on my concerns, not on my opinions.

What I want to say is that I specifically chose *Casablanca* for that recent evening. I knew I would be alone; Sallie had been vague about some kind of outing with May. I suspected the truth was that they had an urge to look at their Biographs on VieGie, to revisit highlights of their respective, previously unshared married lives. So I watched that old movie from the absolute beginning, including the context setting voice-over and the frenetic street scenes that few would recall. I relived Ilsa and Rick's Paris romance. The upwelling of familiarity during the gratuitous insertion of Sam's "Knock on Wood" underscored how long it had been since I watched *Casablanca* from start to finish and how long it had been since Diane left me.

After that exchange of gunfire in front of the hangar, Strasser turns back to the telephone. "Stop the plane," he barks into the handset. "This is Major Strasser! Stop the Lisbon plane, I said!"

The scratchy audible on the other end is garbled, deliberately impossible to understand, per ancient cinematic technique.

"Then shoot it down," he insists. "How many?" he asks without inflection, as if a census worker. "Can't be helped." There's the slightest pause before his imperious repetition, "Shoot it down.... Yes.... On MY authority!!!"

He is quite still, like a statue. Only his voice reveals his fury. His back is to the camera, so it's cinematically up to us, the audience, to recall his prior stony, Teutonic visage. A clever and very effective evocation. The camera then tracks back, to capture the passive triad at the front of the hanger. Strasser and Renault are apart but each intent on the white gloom obscuring the runway beyond. Rick is a nonspecific crumpled heap to one side. We are shown the oh-so-apparent scale model aircraft being lifted into painted clouds. The enhanced roar of engines is to confirm that it has reached the end of the airstrip and is attempting to climb. There are staccato burps of anti-aircraft guns in the distance. We see the corresponding flashes over the pairs' shoulders as they stare through the mist. Then come the whine, the crumply-crunch of a crash, and a faint red glow. It happens quickly, too quickly. Time and distance are improbably compressed in the damp, cinematographic fog. The reality that sound travels at a tiny fraction of the speed of light is ignored. It would complicate the scene's continuity. Hah!

I wonder if it seemed equally stagy to those in our library during that maniacal episode decades ago? I could easily have been there. Being in my office, preoccupied with Diane's worsening state, I didn't attempt to decipher the sharp sounds. My casual thought was that an inconsiderate repair crew was to blame. Regrettably, the noise had a different, macabre origin, but I hadn't paid serious attention to any of it, neither to the day's events nor to the subsequent exegesis. Diane's condition had progressed from fear and discomfort to stubborn pain – biting, tortuous pain. That was my focus; that's what occupied my attention on that day, as on many days thereafter. I never took pictures of the broken glass and stains, as did many. Days later, I did glance at the bullet holes and the superficial repairs, but only because, having heard the sounds, I could then put them in context. Half a dozen or so real people died on that almost-summer day in twenty thirteen at Santa Monica College, including the camo-garbed gunman – who shouldn't have been counted among the victims, in my opinion, neither then nor in the many similar incidents thereafter. Begrudgingly, I came to accept the elevated security, the enhanced police presence, which, after not so many years, was the norm throughout Santa Monica.

    Early that next year, when her pain bloomed, when it dominated even fear, Diane went to the hospital for the last time. The drip made her comfortable and less aware. Both were good. My feelings when Diane died were unblemished by guilt; I had used that up long before. Alone in the house, I could work on lectures or add to my notes for The Book. I could exercise, read, compile lists, or even just stare at the video screen. I was unconstrained. I could explore Venice or Westwood, wander Rodeo Drive in Beverly Hills, or amble along Melrose. I could again speed recklessly through Mulholland's curves or drive north along the coast. No longer was there the need to pretend to be unaware. It was a relief to let everything go, to have dark memories freed from daily reinforcement and begin their inevitable fade. But the silence of the house eventually was itself too much of a presence. Overt loneliness arrived after the utility of routine waned and the apprehension of life as an end more than a means took hold. Sunset Vistas then was the logical choice as well as projected necessity.

    What remains in cognitive imagery is all that I have of Diane with any movement to it. There are photos and slides, of course, somewhere in the closet or a drawer. But I never thought to make a video recording when she was healthy and vibrant; it seemed ghoulish to start later. Her cancer, her dying, had come before the virtual visitation afforded by VieGie, even before Biograph. Neither she nor I had to start on any of it. With the full spectrum of total recall technology now in place and virtually obligatory, gradual metamorphosis of the specific to disjointed, enhanced ephemeral impression is no longer the rule. The past no longer must grow dim. That stabilization has its obvious positive aspects. It's also, unfortunately and in equal measure, a negative, as any with painful memories will attest.

    Do I regret my omission? Possibly. Probably. To relive through faithful images is often to be welcomed. On the other hand, severing the bonds between present and past, which requires forgetting, can be a good thing. The young, the new generations – those totally immersed in games, in trivial amusements and vacuous social probing, in the created truths of prepared news, in the past altered for cause – have less need to detach their memories. The coupling will one day be used against them, as so it is for us, the elderly, through VieGie. Only, we grasp the difference.

Casablanca.

Right.

Of course.

That's what I mean to focus on.

Throughout those final frames after dispatching Rick, Strasser's face remains averted. Yet, his malevolence is palpable. Presumably staring into the fog at what he has unleashed, he's defined yet motionless, still, like a photograph, and quite in contrast to the mobile Renault. The latter walks into the hangar, picks up a prominently displayed bottle of Vichy brand mineral water, then turns partially toward the camera as he uncaps it. With it gripped in his left fist, he inexplicably examines it from the backside. He makes no move to turn the bottle so he can see the label, nor can one imagine the contortion that would enable him to do so. He pours, then evidently reconsiders and drops the bottle into the trash, mutely reasserting his role as flexible opportunist. It's an unnecessary piece of business, dramatically inconsequential, some would judge, but an easily deciphered visual pun.

That's what life is, after all – a host of small moments strung out on a sparse lattice of major events. Those dramatic happenings are the armature, the framework, not the final, sculpted piece of work.

Anyway, at the hangar, the camera's field of view narrows. We see Renault's arm lift another bottle, from beneath the table, with which to fill his glass. The camera moves in further, with the full screen Evian label remaining in focus as the final scene fades.

The film seemed fresh, almost new, as befits watching a classic. But the evoked emotions of loss weren't as I wanted. It would've been far more satisfying if Rick had won the duel with Strasser, if Laszlo and Ilsa had successfully flown off to continue his struggle. Rick and Renault should then have bonded, as the latter's adaptability and Hollywood's liberal, wartime dramaturgic logic would have allowed, to fight the Germans or to move on to some other adventure.

Apparently I've grown overtly romantic as I've aged. The world as it is has little room for either. The actual ending of the film was regrettably au courant but not at all what my selective sentimentality would have preferred.

It's frustrating to so confidently feel a truth yet be unable to prove it, even to oneself. I can't escape the feeling that *Casablanca* wasn't as it should be. What I experienced didn't affect me as I had anticipated. It seemed to have changed.

Or is it I?

CHAPTER 2

ANTICIPATIONS OF SUNDAE NITE

High up in Sunset Vistas' neglected stairwell, at the landing midway to Two, the early evening sun brightens the window and illuminates aging walls. The irregular surface of the failing and repeatedly repaired plaster presents vague yet familiar patterns. Barnard peers upward to discern the identity of the ambiguous gestalts, which are sensitive to the quality of the incident light as well to as to his mood. Overlain is the silhouette of the palm outside. Restless in the onshore breeze, it contorts across and accentuates the three-dimensional space defined by the intersecting planes. Lowering his gaze and rejecting the railing, he starts his patient climb. He prefers to use these relatively unused stairs. Few of the staff go this way, the other residents never. Going down is easy. Going up takes effort of will as well as of muscle.

One. Two.

The stairwell's state is indicative of the general condition of the residence – ageing and less than it was – and analogous to Barnard's often revisited perception of himself. Sunset Vistas used to be a fine, well maintained building. Located in the largely residential neighborhood of Ocean Park, south of the commercial core of Santa Monica, it is close to Lincoln Boulevard and only a short bus ride from the beach. His one bedroom apartment is larger than many – "... with ocean view!" He had judged it was worth the extra cost, even though the rise to the west, the final high contour before a steady decline to shoreline, blocks any decent such view. He has to lean in close and look hard right, astarboard, to catch even a sliver of the Pacific Ocean's gray-blue through gaps in the buildings beyond. Late at night, when alone, he will often strain to hear the crude rhythm of the surf. Ocean view? He had accepted that as part of the usual coastal city hyperbole. The sea's proximity was never an issue. He can go to it, as before. It is not necessary for it to come to him.

With the addition of the upper floors, the Sunset Vistas building is comparatively tall. Local objections to the expansion had been anticipated and were quickly overcome in the Council. The justifications were obvious, namely, the rapid influx of retirees and the generous promises of subsidies. Climate, local amenities, and facilities – leisure as much as supportive – have long made Southern California a prime destination for financially secure seniors. The tide of their patiently acquired and carefully hoarded dollars was most welcome, especially so now, in view of the ongoing financial mess stemming from the recent Big Crash. Having arrived at Sunset Vistas two years before that debacle, Barnard had sold early, bought early. It could not be helped. Size, location, and circumstance prompted his quick decision. He had not been concerned with why the fourth floor apartment had come up for sale. It was available and fit with his plans.

Plans, indeed.

His two rooms, plus kitchen and bath, provide adequate personal space. He has the late afternoon sun. He has tall palms, whose outspread fronds partially shield him from the street below. What view he has is virtually unimpeded. Many, particular those on lower floors, must accept the intrusions of foliage or nearby buildings. Those on the topmost floors are, of course, beyond caring.

Three. Four. Five.

Barnard cannot remember them painting this stairwell except in necessitated dabs. The flaws, the discolored patches, the faint aura of disuse, are hard to ignore in the confinement of its bare walls. Unable to dismiss them, he makes of them a game. The six years since his arrival is too short a span for him to have conformed totally to the lethargic unconcern that lubricates time at Sunset Vistas. This late afternoon he scans the forms suggested by the slant light on variegated, patched plaster. There is the turtle. Higher up are the gulls in flight. The textured blotches speak to him much as had the gleefully anticipated glass cups of pudding Grandma Jane, his Jane-Jane, used to prepare for special occasions, like a birthday, a dinner guest, or a favorable school report. Barnard always had a nimble imagination. It had been fine adventure to assign notable significance to the constellations of blebs scattered in the thickened, rubbery film over moist chocolate. The puddings came less often after his father died. He had just turned seven.

Barnard pulls his head back as he progresses. His neck muscles tighten, impede his attempt to see the monkey head above and just to the left of the high, wide window at the intermediate. The low sun glints off the crackle in the glass and hurts his eyes.

Easier seen coming down. Six. Seven.

Sunset Vistas, even this original portion, is not very old; nothing is truly old in Southern California. But paint lifts easily in coastal climates, especially when the surface is not properly prepared. To slather new over unstable old is evidence of short-term, reactive thinking, not planning. It is frequently an assigned task devoid of substantial benefit. Barnard had often used similar examples of the misapplication of economic resources to enliven his lectures.

The Laura Graese Memorial Activities Room is on Three. The two flights up are about right for Barnard – tiring without being exhausting. The climb is not merely doable, it is proof. Occasionally he takes two steps at a time as further test. Professor Barnard Francis Cordner sets a fair pace but goes slower today than usual, to avoid eliciting an unwelcome sheen of sweat. Summer has arrived. And there is his suit. Wool – even light, fine wool – is a poor choice for this time of year. Still, there was no point in changing after Aunt Laura's memorial service. A distant memory tantalizes him.

"Wool is never right in Southern California."
Who? Mother said? Diane?

His suit is dark gray and faintly striped, perfect for the funeral. He had set aside the double-breasted relic, along with a few other dressy items, when stuffing boxes destined for Goodwill. He has no cause to regret that wholesale donation, although sometimes he does.

Professor Cordner smooths his tie as he continues his climb. It is, for a host of reasons, time to forgo wearing suits, although he has enjoyed the momentarily reconstituted formality.

Italian?

Barnard had attended the service for Aunt Laura less from kinship than because she was an often recognized "good friend" of Sunset Vistas. Their personal relationship was not close enough to engender sadness at her overdue passing, only chagrin. His long gone younger brother, Anders, was the one she favored. Therefore, Barnard felt no sense of loss during the service. Nor, he knows, will he ever. Theirs was not that kind of family. What he felt was a renewed appreciation of time having past, of its inevitability. It has been a long time since he attended any funeral, which, to a degree, is odd considering his seniority. People seem to move away, to Relocate before they pass on. He has had scant need and few opportunities to offer goodbyes, silently or otherwise.

Eight. Ten.

He had not mingled earlier, a choice made easier by the large number of attendees. The few women were seated; the men mostly stood throughout the service. All were attentive. It was a most proper assembly. The red and gold cloth, edged in black, that sheltered the cradled coffin, was ostentatious, he decided. And the late afternoon June sun was warmer than he had expected. He could have done without that and rued his lack of head covering. At a Jewish internment there would at least have been one of those silky, blue and white beanies to put on. Even a black one would have helped.

Frances, Barnard's mother and Aunt Laura's substantially older blood sister, had died twenty years or so before, at the end of the noughties, shortly after the bursting of the post-millennial credit bubble. Barnard was not unique, in light of the excesses of the dot com fiasco that preceded it, to foresee that. He had presented many anticipatory talks to postprandial collections of often preoccupied businessmen. Lacking the tools and talent for practical investment, being immersed in purely academic economics, he derived little personal financial benefit from his craft. He ended that first decade with as much in his conservative investment account as he had at the start, which was not as much as he would have liked. Aunt Laura, he could not help but occasionally consider, had considerable means. She could easily have moved into Sunset Vistas – or Eastern Star, as it was then called – as had Barnard's mother. Childless and already a donor, she had been courted to do so. Barnard was always glad she had declined. That would have been difficult for him. Ten years old when he was born, Laura Graese always struck him as a cold, overbearing sister more than an aunt. He was grateful for her absence.

The large crowd that had showed up to acknowledge her passing was a surprise. Barnard experienced a pinch of irrational jealousy as he scanned their motionless backs. He recognized none of them. The few relatives he knew of were distant, all female, and never part of his life.

Barnard reviewed that during the tedious return drive. Feeling submerged in the confines of the van, the constricted sight lines created a cocoon of perspective that seemed to favor introspection. He ran his fingers over the hard edges of the key in his trouser pocket and attempted to conjure up, in the confusion on the glass of the adjacent window, long absent faces, the ghosts of his family.

She was active at the Rotary well past her sixties. "Are they all from that?" he had asked of the empty seat next to him. Could the turnout derive from the Graese Dairy Foundation having been a responsive donor? Santa Monica is neither so big nor so wealthy. A little went a long way. While he receives oblique benefit from being her nephew – an odd situation given their closeness in age and mutual disregard – he is not gratified by the recognition. He did not, does not share her views, which, while becalmed compared to those of the ultraconservative core of his departed immediate relatives and, presumably, those distant who are still alive, are on a similar course.

Happily A.G., his grandfather Adolph Graese, is long gone. Barnard, in his independent, adult years, came to view the man as a presumptuous, fast-talking toad. Formal yet glib – "It All Starts With Milk" was his motivational slogan – and exceedingly insistent, he was a lightning rod while he lived. As a "kleiner Fisch," i.e., little fish, his views served both sides, were as useful for the ascendant Orange County far right to elevate as for the beachy, intellectual left to castigate. "All politics is local," the politician Tip O'Neill was supposed to have said, over a half a century ago. Barnard finds it hard to reconcile the Graese family's financial generosity, a family of which he is a part, with its retrograde, narcissistic conservatism, of which he is appalled.

Eleven. Twelve.

* * * * *

His Aunt Laura's funeral service went on longer than he had expected. Her local political interests, with financial support overshadowing leadership, had otherwise mirrored those of the Graese patriarch. There was much to cover and the remembrance, presented by a tall, sepulchral sort, spared no detail. He, a funeral home staff member or nondenominational clergyman Barnard presumed, was exceedingly smooth. His remarks obviously were prepared with diligence. Not ensnared in their flow, Barnard attempted to visualize the array of Old World hors d'oeuvres that he expected to be provided.

The rustle of stiff clothes and the clatter of chairs in haphazard contact alerted him, signaled the end of the graveside service. It would have made good sense to return directly to his apartment, then to change for a slow walk down to Lohren's OP Café for a pressbrot. He could have slipped aboard the shuttle before it left. Instead, his hesitation left him to choose between idly waiting for its return or joining the others at the post hoc reception. This lapse was not totally by accident. There was the anticipated buffet.

He followed the group inside, where his supposition proved to be correct. Undisturbed, he cruised past the array of offerings. There were tangy stuffed pastries. Thick smears of egg salad between white bread squares, their crusts cut away, sat near comparables on pumpernickel. The hard balls of meat and egg yolk on toothpicks, in thick brown sauce, went well with the cheesy potatoes and onion. The herring had smelled off. Whether from being set out too early or from being too old was relevant but undecidable. He declined it. He likewise should have passed over the heavily smoked braunschweiger. There was no wine, unfortunately, only tall glasses of beer. A chilled Auslese would have been refreshing after the too bright coastal sun. He found an isolated seat and tucked one napkin securely under his plate, holding the other lightly as he ate. Fork rising, Barnard briefly caught the eye of the eulogist.

On a second pass he found cherry cake, precut and plated, crusty sweet rolls and a tray heaped with cubes of Allgäuer Emmentaler, Butterkäse, and Bergkäse. There was even a server laden with quarken Milbenkäse, which sat virtually untouched. Barnard had not tasted the like in years. The ethnicity of the offerings was intuitively familiar to Barnard, a revivification of early family gatherings. He smiled ruefully at the thought of a glass of Trockenbeerenauslese.

The vaguely unkempt, pudgy individual who approached Barnard was totally unfamiliar.

"Aren't you Bernard Cordner?" the man asked.

"Barnard, actually. Barnard Cordner," he responded. "Have we met?" he then politely inquired, feeling at a disadvantage by not having the faintest idea of the person's identity. He wiped his hands carefully with the extra napkin before standing.

"Oh, I'd be surprised if you could recall me. We met at your father's funeral. We were both kids. Do you recall?"

2 -- ANTICIPATIONS OF SUNDAE NITE   23

"Sorry, I really don't. You are?" Barnard put to him as he searched for a convenient spot for his empty plate.

"Kurtman Fowler, Barnard. Good to see you."

"Right. Nice to meet you. Or see you again, as you say," Barnard said civilly as they shook hands, putting on a smile to ease his awkwardness.

"Hmm. Yes. Amazing. My father, he was Kurtman Fowler, too. I'm actually Kurtman Junior. My father knew your grandpa, Adolph Graese. He had me sit with you while the two of them talked. It stuck with me. The memory, I mean. I was so glad to be next to someone my own size. It was my first. My first funeral, I mean. Do you recall?"

"Right, I do. I believe I do," Barnard lied. "It was a long time ago." Barnard's thoughts were focused elsewhere at the time.

"It was. It must be over seventy years."

Barnard made no attempt to refine the estimate.

"You live here, in Santa Monica?" he asked.

"Oh, no," Fowler laughed. "In Orange. I came up with Doctor Weill, in his car," he said, indicating the eulogist, who was then in conversation with two trim, young men. "I work for him. With him, I mean. In Orange," he repeated, as if that were somehow significant.

"I see," was all Barnard offered back. His unmasked indifference brought faint color to his cheeks, which Fowler seemed not to notice.

They talked superficially about who did what, about prior and present activities, about children, family, their respective wheres and whens. The usual forgettable details were passed between them, making the long interval since their prior meeting no less a gulf.

"I'm sorry to hear that, Barnard," Fowler stated softly at one point. "She was too young, I mean."

It was then that Barnard felt that the encounter had dragged on too long.

"Well, uh, it's been nice to visit. Perhaps another time, a happier occasion," he suggested to the still unrecalled Kurtman Fowler, with no conception of his reference. Nevertheless, he offered his hand before turning to leave.

"Yes, Barnard. I'm sure we will." Fowler matched him step for step as they exited the hall. "Busy place," he remarked casually as they stood on the drive at the edge of the cemetery's pleasant grounds. They lo-'
the expanse of grass to another well-attended finali

ceremonial bobbing of covered heads under the distinctive blue and white canopy.

"There another one made good. Not a moment too soon. They should have their own separate section. Or none at all." Fowler patted his thighs, as if he were searching. "Nonsense. Them pretending to pray, I mean," he appended through tight lips. "They killed the only real Savior. Who could they be praying to?" He pivoted sharply toward Barnard. "Would you like to be taken somewhere?"

"No. But thank you," Barnard replied, revealing no reaction to Fowler's observation and poor grammar. "It'll take me to, uh, take me home." Barnard lifted his hand in the direction of the electric powered shuttle that was by then waiting a few dozen yards down the drive. "Thank you, for the offer," he disjointedly repeated to Fowler, relieved there was no need and careful not to let that show.

Sated, by more than the food, he was happy to finally get into the van. No crowd will show up for him, not for Professor Cordner, Retired, he was prompted to consider. No, he decided. For his service they will find a priest to speak at excessive length about one whom he knows little of, of one whom little need be known.

His room felt particularly empty after the funeral. Barnard did not attend to the sound or the images of the Internet screen that immediately came to life as he entered.

"Hello, Pro Fessor Cord Ner," the ever alert Daedalus Man offered before Barnard could command down the volume. He waved up the menu and pointed to mute, preferring not to speak. He gave in to predictable ennui and moved a chair closer to the window so as to sniff the breeze, to feel it. He wanted to bring life, the world, into his room. He strained for the distant murmur of the surf and tried to taste salty mist, opening his mouth to breathe. Some, in other buildings, had foreshortened patios that hung out over drives or walkways. Staring into the vague distance, he envied them and simultaneously wished he had taken a second helping of the Nobel Blue, the Edelpilzkäse, on a toasted crisp. It was a crumbly mess, it seemed at the time, but so tasty. He was unsure when he would possibly see the like again.

Sitting motionless, breathing slowly through pursed lips, Barnard watched gulls careen across, anticipating, as did he, the evening. From left or right each came, then disappeared to right or left. Occasionally he heard a harsh squawk or saw aimless ejecta arc downward. Fragments of ancient history enlivened his overextended, contemplative interlude. Are they identical? the fleet birds finally caused him to pose. Or is it the same lone gull, changing its identity by the going back and forth, like Feynman's electron/positron duplicity via the transgressing of time's arrow? He chose not to attempt an answer as he carried the straight-backed chair back to the kitchen. The sight of his reflection in the glass panel of a cabinet door prompted him to run his hand lightly over his thinning hair.

* * * * *

Thirteen. Skip Fourteen.
Barnard sees obscure shapes dart by the window that comes to eye level as he steps onto the landing at the turn of the stairs. More birds. He pauses. This afternoon's service has eased Barnard into thoughts of religion, something which is generally not high, to say the least, on his agenda. However, the intrusions of ubiquitous casual theology have become more frequent. Bad times do that. The once arguable, now supposedly proven inadequacies of secular government have provided the politicized Theocrats their long sought opportunity to legitimize a strong leadership position for themselves.

There it is. Born again ascension of inflexible tradition. The New Motto: "In God's Plan We Trust." One controlling concept cast out, recast to rule again. Re-creation of what's been destroyed. Resurrection? The Dance? No, not that. Circus? Right, that fits. A circus of back and forth. Different, yet the same. Like it, earlier, the gull.

Hah.

Myths and fables. Best focus on the real, not on nondemonstrable misdirections. On the social symphony of history, on its recurring rhythms and rhyming variations. Maybe The Dance is right after all. Step and ...

His thoughts fade and he attends to the stairs. There are the pressures of circumstance that must be addressed. Barnard is, as are most, pinched between the jaws of a patient, inexorable vise. On one side are the lawyer politicians, content to prove anything. On the other are the anointed guides, insistent that proof is not required. Their union of convenience has only exacerbated his dilemma.

Able to explain everything. Sure sign of understanding nothing.

Jacket, bad idea.

2 - - ANTICIPATIONS OF SUNDAE NITE

He pulls on the middle button of his coat. The desire to retain some vestige of his former persona has compelled retention of the expensive suit. Barnard always had shunned lightweight businessmen's uniforms. Never wore seersucker or polyester. Neither had he accepted that feeble caricature of academe, the sports coat with patched elbows, emphatically rejecting Diane's well-intentioned suggestion that they were "smart." In addition, there is vanity. A well-crafted jacket's structured shape serves to hide the expanding flaws of his body. This evening he spies his corrugated shadow on the stairs ahead and stiffens, trying to forget the recent hours in favor of an anticipation of this Sundae Nite. He feels the sun on his nape and takes a measured breath before continuing his ascent, vaguely aware of the musty stagnation of the stairwell. A distressed corner of Sunset Vistas, true enough. Still, dust can sparkle pleasantly when light streams in at a shallow angle.

One. Two.

He could have gone the other way, used the elevator at the main end of the building, except that he had wanted to check the package room, one last time before the weekend, for a half-expected token. Barnard envisions four potential signees, having long ago given up on his daughter-in-law as well as the inconsequential, if still extant, idiot that his daughter, Katie, has hooked up with. Barnard – Father and Grandfather – accepts that a remembrance may yet come via net-mail. In his view, that would be a poor substitute for a physical card. One can never identify who conceived what by the impersonal fonts of those ephemeral missives: old or young; personal or generic; actual or surrogate. Still, he had hoped.

Three. Four. Five.
Not one goddamned card.

It vaguely recurs to him to establish to which side the blame for the perceived neglect should fall.

A faint sense of effort arises to join the warmth in displacing his mild disappointment. Yes, he should have exercised today. Except that there were the funeral and the buffet. Yes, he should have made use of the Physical Therapy room or taken a walk somewhere. Tomorrow possibly, he muses as he paces upwards. This early evening Barnard is putting the stairs to good use.

He ponders what movie he and Sallie might watch later tonight. She prefers the domestic new offerings or the foreign site feeds. Barnard is less so inclined, preferring classic films or short, action shows, videos with, at least, logical and internally consistent plots in direct English. He could watch a John Wayne western or one of Douglas's over the top, macho trivialities a dozen times – a Bogey movie, most made before he was born, two dozen. Any scheduled appearance of Raft, Robinson, or Cagney, all of whom also predate him, is always tempting. Barnard's action contemporaries were Eastwood, Connery, Harrison Ford. Yes, were. All are gone. Crowe is alive, as were De Niro and Damon until very recently. Gibson, incredibly, is still gnashing about, although he is no longer fun to laugh at. He is too much of the moment, his prejudices no longer newsworthy.

An A.G. clone, sans intellect.

Six. Eight.

They managed at least a passably satisfactory conclusion in those older films. No longer. So much now is contrived for temporary effect and rapid wrap-up. While some from recent decades are worth watching, Barnard has retained few in memory for replay on sleepless nights. Occasionally a domestically produced drama, something beyond shaped noise, comes along with sufficient valid affect to counterbalance effect. There was that fine film of De Niro's, a virtual bookend to pair with his early work. He died before they could finish shooting. They had to doctor the close-ups and reshoot scenes with surrogates to get it wrapped – quite paltry tasks compared with the current craft of manipulations designed precisely to render seamless the transitions to fantasy.

Nine. Ten.

Barnard's age-imposed high frequency hearing loss complicates his enjoyment of video fare. Sallie is always urging him to lower the volume and to use subtitles instead. He prefers to make it louder. Even on the large screen in his living room, the text is often too small when at the bottom and can be intrusive when oversized. Either way, flicking his eyes up and down disrupts image flow, compromises the illusion of depth. Last night it was Sallie's turn to choose. Her selection seemed odd to Barnard, since the plot summary indicated it would be a thriller. On the other hand, the dialogue was relatively clear for an import, which was helpful. Separating dialogue from background can be difficult for Barnard when sound effects and music are pumped up to excite the viewer in real time. The newer dubbing technology helps. It dynamically shapes mouth parts to fit the translated speech, making it easier for the viewer to stay in synch. The downside is that intonation often goes flat, which can obscure the character's intent. Barnard has to pay close attention.

Eleven. Twelve. Thirteen.

The late sun shifts to his face as he yaws across the second floor landing. His in-breath carries that familiar hint of the sea at this, the midpoint of his climb. He wets his lips, seeking more

There are heavy fire doors at each floor except for one. The windows opposite – none of which open – are invariably high over the stairs. The wire in their reinforced glass adds to the complexity of the cast shadows, transforms the mundane into evocative imagery. Barnard looks up at the black TWO stenciled above the door and catches a glimpse of Professor Cordner in its small, square view port. He adjusts his posture before continuing the familiar patient ascent.

One. Two.

The elevator would have been slow, its bare steel walls too constricting, its redolence too definitive. He senses a comparable confinement as his sleeve brushes a wall, but there are additional reasons why Barnard is using the stairs this evening.

"Oopf!"

He glances down, aware that he will find no excuse for the misstep. It is another annoying sign. How fine it had been to stride firmly without consciously attending, for him to sense and for his limbs to anticipate nearby objects, to feel secure no matter what was underfoot. His eyes dart down and up to confirm his next few steps.

Four. Five.

He licks his lips again. Their dryness is seasoned with recent indulgences. Paradoxically, he is aware of a vague, internal emptiness that accentuates the closeness of the stairwell.

Six. Seven. Eight. Snack in the Activity Room.

AC turned up. No surprise. Bad, though. Too warm. It's Summer.

Nine. Ten. Eleven.

Not tired.

Twelve, fourteen.

"Pfffhhhhh."

He has successfully skipped another step and therefore straightens more energetically when he finishes the penultimate internalized count, as if someone were watching, evaluating. At this turn midway to Three, the late sun's strength is more evident since the high window is not troubled by waving fronds. Its aligned palm – a replacement after some obscure disease claimed its precursor the year Barnard moved in – is not yet as stately as its adjacent cousins.

Taller soon. Everything in its time.
One. Two.

Professor Cordner is aware he will be late. He looks up and to his left as he begins the final segment of his climb. Three is the only floor to have an exit door of thick glass instead of painted metal. Barnard has never settled on exactly why, other than possibly because this corridor was reconfigured for office suites during the conversion. Even if unused, the architecturally decorous door can be seen along its entire length. As he once had taught, such budgetary concessions often serve as valid adjuncts for reinforcing status and hierarchy.

Three. Four. Five.

The glint from the door's massive contemporary pull catches his eye. Then, through the lower portion of glass, the unoccupied hallway, lined with opposing pairs of numbered entries, comes into view. It is a progressive periscopic view, one of unpromising linear gray, not the broad expanse of a sea under bright sun or moon as in a well-executed breach. It is an unplanned allusion to another aspect of the course of his years. There had been many doors from which to choose; what lay behind each subject to discovery, rarely prediction.

Make a Deal. Monty's paradox.
Six.

Secure in his climb, hand lightly on the rail, Barnard's imaginings dive into the past, then link with it and pull it forward. Diane died nearly eleven years ago. It was thereafter left to his son Kyle to remember special occasions. He has done so five times out of eleven, counting the last. For the third or fourth time this past week, Barnard reviews the success ratio. A Father's Day card this time would make it precisely fifty-fifty. Fifty percent, nearly the same as the second of the Dow's early post-millennial declines, that of 2008-2009. The Dow's prior Y2K collapse was nearly thirty-three percent. One third. One over three; its decimal sequence never ends.

Incidental numerology. Coincidental.
Seven. Eight.
Damn eCards.

"Pffhhhhh," he sighs.

When you have grown up with tangible greetings, digital counterparts lack the same impact. Barnard's rational mind grasps that the latter are equally sincere and, therefore, deserve to be equally well received. Nevertheless, he regrets the lack of intimacy. He remembers how it was to hold a gatefold card, to feel the reality of a physical entity to take to the light to decipher its scraggly crawl, something to put in a drawer for later review, or to share with someone and bask in their response. These are the lost options of a demonstrably different time.

"Certainly not!" most would rejoin. "It's an improvement, a replacement of one mode of personal contact with a more functional equivalent."

Right.
Nine, ten. Eleven.

He increases his pace up the last few steps to the third floor landing so as to prove that he is not declining, that he will not accept it. Not Barnard, who presents so well at 79 as to be taken for a young 65.

Own fault. Four, including the two grandkids. Five, with Katie. But no card. Autogenerated, even, would be okay.

Twelve. Thirteen.
Three.

It is quiet, except for Barnard's prolonged exhale and the echoes of his final, scratchy steps. He pauses on the landing. Through the glass, the hallway ahead is dimmer than where he stands. The light from behind commingles with his floating image. The ungenerous reflection is that of a compact frame, beneath a long face with thin, pursed lips. The high forehead, straight thin nose, faintly olive skin, and irregularly shaven chin hint of Italian more than his factual German-Irish stock. Noting the dour expression, Barnard elongates his lips into pleasant neutrality then lifts his gaze to the gray tousle above. He purposefully pulls his head back and straightens as he runs one hand over his hair.

He looks up at the underside of the stairs that lead to Four. In the light of the half-silvered, solid-state fixture that washes the surfaces above him, the strands of a cobweb – stringy glints in the corner of the seemingly lifeless space – scintillate as they waft upward in the wake of his arrival. It does seem bigger today. He waits. Yes, it sags down. Change often comes too slowly to be noticed easily. There must be tests; there must be patience. Barnard searches for the sparkle of tiny reflective eyes.

Not even a goddamned eCard.

His suit, his anticipation, his straightening his shoulders – all because of Sundae Nite. Barnard reaches for the brushed nickel handle. He notices staining on the glass and makes a mental note to wash his hands before going into the activities room.

"Ta Ta Ta Thum"

His hand and his head jerk back at the sharp intrusion of Beethoven's Vee. The musical phrase repeats as Barnard explores his pockets until, at last, he finds and can retrieve his cellphone. No longer muffled, the alerting tone resonates sharply in the constricted space. He squeezes the unit between thumb and pointer finger.

"Don't forget!" he is brightly chided, "Two For One at your Wyse's Bite Box!!! It's always nearby. Get yours for dinner today. Eat Wyse!"

A pair of pretentiously thick and juicy soy burgers share the screen with the familiar winking owl. An overlay of a map of central Santa Monica appears under the logo, a bright red arrow designating the intersection of Wilshire and Fifth.

Never adverts for chocolates. It knows.

Instant on would be fine, was it not for the commercials and the implicit oversight. Barnard vows to reset the unit to SILENT/VIBRATE, immediately after he takes this call. He taps the ANSWER prompt. The premeditated solicitation is replaced by a familiar, momentarily distorted image. Sallie's petite nose looms large.

"Hi, Sallie. Don't hold it so close."

"You said you were coming, Babe," she replies with a falling tone, a question reconfigured.

"Later," he offers suggestively.

She makes no reply, merely tilts her head and looks to one side, as she typically does when he misdirects in that way.

Barnard grins at her.

"I am, Sallie. I'm on my way," he then affirms, by way of a proper answer to her observation. "I wanted to check if Kyle or the kids had sent anything. For Father's Day. Katie even. Right? At least one, uh, at least one of them could have, uh ..."

Sallie's face recedes to normal perspective. "Kyle's busy and the kids are already in camp," she says. "Aren't they?"

"I guess so. I certainly didn't expect anything from Katie and her Freaky Foeman."

"Don't call him that. You might let it slip one day."

"I won't," he says. "Besides, I don't gara."

"What?"

"I don't give a, uh, I don't care. I try not to let on that he exists. Witless fascist."

"Stop, Babe. And don't fret over a card. Kyle will VieGie tomorrow, or Sunday, and put the kids on. Maybe Katie'll call."

Barnard says nothing.

"You're not in your apartment, are you? Doesn't seem like it. What did you do for dinner? Did you fall asleep on your couch again?" Sallie puts to him in a rush.

"No, I didn't nap or anything. There was a reception after, uh, after the funeral service. My Aunt Laura's memorial. Right? I told you.... Yes I did." Sallie has tilted her head as if to contradict him. "I'm sure I did," he therefore insists. "It was hot outside there. Knew it would be. I was roasting, in my suit, until I went inside. They put out a table of old country snacks that I, uh, that I hadn't, uh ..."

"So why did you go?" Sallie inserts into the pause, as if the answer was not obvious.

He makes no reply. Instead he lifts his eyes in the direction of the web in the high corner.

"I should have called earlier," Sallie says, so as to modify the sense of her admonition. "Did you go to PT? Use the elliptical?"

"No. I was at Aunt Laura's do most of the afternoon. I got back too late. Took the stairs instead."

"Instead of exercise? Well, that's not going to get your heart rate up."

"Right, Sal. You're right. Saaay, I've got a great idea for that. What do you think?" he coos, lowering his pitch for feigned seduction.

"Stop."

"I'm on my way, Sweet," Barnard relents. "I'm already at the far end of the hallway."

There is no sense describing his afternoon to her – the crowd, the food, the hot afternoon and cold encounter. They will have plenty of time to talk after.

"Is there much of a crowd?" he asks.

"The same. Well, maybe fewer, I guess," she informs him.

Studying the web above, he starts to describe a house spider's tenacity and how remarkable that is, in light of the meager prey creeping or flying in that space. Sallie's profile reveals her attention is directed elsewhere, toward May, who is chattering in the background. Both laugh at some hidden piece of business. Barnard wonders what.

"Turn your head, Babe," Sallie requests when she reattaches to their conversation.

"Hmm? What?" he replies, searching for context.

"Turn your head. I want to see."

He moves the unit away, holding it at arm's length, so she can get a proper view of what he realizes is her concern. He cannot restrain his lips from tightening and feels a light flush invade his cheeks.

"Don't be such a fussy. Oh, I can't tell. I'll take a look when you get here. I hope it isn't blunt. Last time it wasn't tapered at all."

"It's fine, Sal. It's what I, uh ..."

Again, he chooses not to finish.

"Don't do that. It's so much neater when she tapers the back." Her face shrinks further. "Such tiny things annoy you, Babe. That's not good. If you would –"

"Wait. I want you to see," he interrupts and lifts his phone up toward the corner.

"What this time? One of your spiders, I'll bet."

"I've seen this one several times. Its web, at least. I haven't seen her. The web is definitely bigger today."

"Looks like a dust bunny to me," Sallie offers without real interest. "I'm going."

"Wait. You can see a tiny white package. There. The round thing, there, in the web. See? That's new. Can you see it?"

"Maybe. I think I can. Hold it closer."

Barnard takes a step forward and raises the cellphone above his head, putting it closer to the irregular shape nestled in the confusion of dusty wisps. To Barnard, the stairwell's illumination and the fading low sun cooperate to reveal a diminutive, shimmering sphere. This is in sharp contrast to the dull curtain of the web in which it rests and which hints of an elusive presence. What he sees, the unit does not capture, however. The angles to his eyes and to its lens are too disparate.

"Hold it steady. I see ... a ... mmm."

In the subsequent silence Barnard imagines her squinting at the image on her own phone.

"That's what these indoor spiders do," he explains. "They don't spin out those pretty, symmetrical lines. They put up this blob. Right? Not a web exactly, a scraggly blob. Then they wait."

"A blob. Okay. I'm glad you're fascinated. Just hurry on, Babe, before the ice cream's gone. It seems that they've cut back. What do you think, May?" Barnard observes her look off to one side. "You should have taken the elevator. It's too hot for you in a suit," she returns to suggest gently. "And don't stand there staring at some spider web. Okay? We'll be at a table. Bye."

He hears May's faint lilt from beyond Sallie.

"Hi, Barnard. We're wait-ting."

Barnard discerns a flash of May then of the ceiling as Sallie lowers her phone. "Bye, Sallie," he says and taps the END button.

The screen obediently presents a picture of Kyle and the grandkids, with an obviously cropped picture of Katie superimposed in one corner. He again appraises the cobweb mass. Constructed, yet formless, it sits in light so changeable that its sheen has already dimmed to a pale spattering of unstable highlights. He sees no hint of its resident, only that tightly wrapped, white bead.

"Must be dinner, saved for later," he says aloud. "It, uh ..."

His voice fades away.

Built that way. Innate. Right. Choice? Plan? No. Joy neither. Just does. Intricate, tiny black machine. Its meal wrapped in white and waiting. Or eggs, maybe. A she.

The spider does what it does because that is the way it is built. Its behavior is defined by diminutive controller networks and the stimuli that drive them. They must appreciate time, because they die. But with action occurring without choice, they have no past, no future. There are just existence, decline, and death.

Barnard pulls upon the curved handle of the door, taking notice of the balanced weight of the thick glass. The less stagnant air that greets him is a pleasant change. His cellphone still gripped and in motion, its abbreviated montage of family photos has been replaced by another tedious advert. It is perfectly silly for Barnard to keep his mobile unit on standby 24/7. He rarely gets important calls. Most are reminders or system notes, revelations of special sales opportunities, or from people he is going to soon see. Except, there is always that chance of a call from Kyle or one of the grandkids. After he confirms the proper locale – the inner pouch of his suit jacket's outside pocket – he lets the unit slip into place. He enters the dimly lit hallway, glad to be finished with the climb and to resume full stride. Again he pulls his head back and squares his shoulders, inwardly ticking off the entries to either side of him as he proceeds: door, door, double door.

One. Two. Three, four.

It is getting harder for Barnard to ignore his calendar age. There are those fleeting reflections of an aging man at a storefront or, much like this evening, in the glass of an entry. There is the inescapable morning visage in the full length mirror on his bathroom door. He more frequently anticipates close examination by others, and so, when on one of his outdoor explorations, he will move smartly and endeavor to project the image of a vigorous, sixty-something male. He will straighten and summon up an echo of his lapsed naval posture. These are vain efforts, truly superfluous, but justified by the inner sense of pride that attends enhanced remembrance. Especially if cued by the presence of others, he will tighten his stomach muscles as if, with a pose, to make up for years spent hunched over keyboards, notepads, and books. So it is on this early summer evening, on this short indoor walk, presumably unwitnessed. Vanity is a sin only when one cannot manage or afford it for oneself.

Appearance can be modulated but not ignored. Barnard perceives that his age is correct, now. He appreciates its associated heaviness, now. The easy walks seem harder, longer. Time itself is a recognized enemy, now. Shifting about on the cold seat in the van after the funeral, he had felt the need to reconfirm that Sundae Nite is today, not the Friday night next. It was not a matter of having forgotten. It was the need to be certain, hence to recalculate. He is confident that he does not forget but cannot deny that he has occasion to question. Realistically, one is as bad as the other.

Test. Tested.
Enough of that.

"Pffffhhhhhh," he exhales heavily.

Today is the second Friday of the month, Sundae Nite at Sunset Vistas. And it is nearing eight o'clock. The treat tables are usually made ready by seven. His vacillation, his indulgence of mobile, disparate images, had yielded to a gray vagueness. He had felt it arrive earlier as he sat looking out. He felt it again as he talked to Sallie a few steps earlier. He feels it now as he cruises along the hallway.

Five. Six.

While his stride is purposeful, he knows that even just a walk around the block earlier would have been wiser than sitting idle. Barnard prefers to accept that The Machine – an elliptical, which is "Easier on the joints ..." – plus his beach walks and exercises on the sand a few times a week, are adequate to keep him in decent shape. He sold his heavy-framed trainer in the move. Fortuitously, a nearly identical apparatus resides in the Physical Therapy (nee exercise) room further down this corridor. It and the other equipment – the treadmills, the low-budget stationary bikes, the floor pads, exercise balls, the dull grey weights – had been deemed obligatory for accreditation of Sunset Vistas, which was a necessary initial step toward receiving the federal subsidies that financed its expansion. No one had expected that insolvency would thereafter push aside benefit as a national legislative planning issue.

*   *   *   *   *

In his school days, Barnard – friends called him BC, Cordy, or Mister Cee, because he rejected Barney – came across an interesting style of solitary exercise. During the 1930's, an ambitious immigrant, Angelo Siciliano, had parlayed his superb physique into a decent living as a model and teacher of body building. Siciliano had the foresight to take on the *nom de faire* Charles Atlas. He used his muscled image to popularize his branded physical improvement regimen, one that pitted muscle against muscle, agonist against antagonist. Anyone could do it with no equipment other than his own body. After that first Great Depression, the consequent war and recovery, his ads still were to be found in comic books, pulp fiction, and periodicals. The before and after graphics – in the quarter page, sometimes full page spreads for the Charles Atlas Dynamic Tension program – usually featured a transformed bean pole boldly knocking down the sand-kicking bully from a previous panel. These provided a feasible fantasy for many maturing adolescents. The name and brief description of the method gave Barnard all the instruction he needed; he never sent in money for a manual.

With immediate family by then consisting of Mother, Aunt Laura, and Grandma Jane-Jane – the much younger Anders did not count – plus two older, paternal second cousins in nursing school who occasionally were invited to Sunday dinner, Barnard felt with particular acuteness a need to be manly. He would strain biceps against opposing triceps, quad against hamstring. Stomach, torso, neck – he worked each in sequence. As his adolescence progressed, the sound of him through the door, his flush afterward, convinced Mother he was being excessive in a way she could not fully attribute to exercise. She would send up the cleaning girl to get him, ample evidence that she was not comfortable with his nascent physical maturity.

Or perhaps, she was.

When out on his own, Barnard came to prefer real exercise – karate in particular – over the static tension workouts. While he enjoyed sailing as a leisurely activity, the explosive workouts at the dojo, the hour or two of formally correct, quick physical movement served a different purpose. The sensei made demands for style, speed, and accuracy that Barnard found satisfying. Soon marriage, the house, professorial duties, and maturing children came to leave him with considerably less available free time. Remiss in his practicing, sparring often left him at the mercy of aggressive opponents. His last formal teacher, an American who had been a pupil of a Japanese master, challenged each in the class to match his own, laboriously attained level of concentration. By then toeing middle age, Barnard virtually fainted one day, had lain on a bench to keep the puke down and, embarrassed, never returned.

Taekwando was not easier but the class was. Even when sparring, it was not as intense. It better suited Barnard's increasingly sedentary lifestyle as an economics professor at Santa Monica College. He even put his son, Kyle, into it for a short while. Practicing his forms, thereby retaining a modicum of skill in martial movements, was worth the encroachment upon Barnard's meager unscheduled hours. Being physically active outside and alone, or in the regimen of an indoor class with others, occupied his mind. Routine, repetition for its own sake, can take the edge off an unpleasant present. It subdued circular rumination. Meditation, had Barnard elected to try that route, might have yielded a similar result.

Later, after Diane died and he was alone in their house on Second Street, informal taekwando groups drew Barnard to the Venice beachfront. He had, by then, stopped sparring at the dojo. He found that repeating the well-practiced exercises on his own was sufficient. Rhythmic, noncompetitive full motion on the sand was kind to his aging joints. It was a different path, a new Tao. When even that became too intense for him, he recalled his adolescence and blended the Asian formalities with Dynamic Tension. He started doing taekwando in slow motion, striking at sluggish phantoms. By blending the choreography of earlier years with a plastic rigidity of arms, legs, and torso, he could work up a satisfactory sweat without excessive strain. No makiwara or elusive opponent was required. The workouts kept his weight stable, which he took as license for having more lunches out than he otherwise should. In addition, he imagined that the retained skills could come in handy one day. Barnard frequently sensed simmering social anger, a forewarning, like distant white caps under a darkening sky, of a coming confrontational milieu. Also, on the beach front when he would stop to rest, there were interesting people to watch on the concrete pathways: young, shapely girls; runners; skaters; strange backwards walkers. Unfortunately, there were also impatient biker pains in the ass who heeded no rules.

People often slowed to watch what they interpreted as angry tai chi. A few even took to imitating his pleasantly aggressive ballet. As with closely cropped hair or tattoos of lightning bolts, as with vandalism couched as protest or acquired naked hate cloaked in convoluted reason, it might someday be commonplace.

* * * * *

Seven. Eight. Nine.

It is dim on this portion of the corridor. The nearby offices are darkened and seemingly empty. Few of those on salary work on the weekend. Besides, so many have been let go. Social Services and Accounting have moved out totally, as has Medical/Nursing. Only semiskilled aides and core operational people are left on-site, in addition to those tasked with bringing in new residents.

Barnard straightens to enhance his stature, his head straining for the ceiling. It is Sundae Nite. He blinks slowly, raises his chin, thinking of the tasty snacks earlier, of having avoided a starchy meal in the dining room, of the treat tables waiting in the Laura Graese Activities Room ahead. Before that terminus, across from the PT's wide, open portal, is Chen's office. Among her other duties, Julie Chen serves as the Arrivals Facilitator. She was Barnard's first official contact with Sunset Vistas as a potential resident. "For ah credy tashun, Doctah Corhdnah," she had said of the equipment in the Physical Therapy room on that walk-through.

She in?

To this day the peculiar intonation of her speech eludes him. Much of what she says, particularly when she is agitated, is lost upon him. While he has heard that cadence, that lilt, elsewhere, he cannot extract its precise locale. His naval service had entailed Atlantic Ocean patrols for the most part, not the Far East. After six years of residence, Barnard still is not sure of the proper pronunciation of her natal given name. It is on her badge and the name plate on her desk, but he has never heard it uttered. Everyone calls her "Julie," so that she has unambiguous given and family names, same as a native born. She is a capable girl, he judges, one of the few regularly here on weekend afternoons, making calls and arranging visits.

* * * * *

Throughout his first visit, Julie Chen's introductory tour, she made up for her unimposing stature by projecting great energy, revealing no hint of playing a well-rehearsed role. Voluble and enthusiastic, she ushered them into her office for an overview of the features of Sunset Vistas: the layout of a typical apartment (there were no accessible vacancies at that moment and residents still were secure from intrusion); the advantages and conveniences; the general and specific facilities. Kyle and his bride tried to project that they were new to these details. It was a solicitous effort. It was also tediously incongruous. Barnard was well aware that Kyle had, in fact, researched the facility quite thoroughly. He knew also that Kyle had revealed to Ms. Chen much about his father, in particular that he took his workouts seriously. Therefore, it was no surprise that, before going downstairs, their first stop was the PT room. As they dutifully admired the equipment, she told of the necessity of the assemblage for accreditation. Making it seem like a casual afterthought, she also had them look in on the Activities Room named in honor of his Aunt Laura.

They were shown the dining room and kitchen on the first floor, then the media and the meeting rooms, the crafts areas. Chen pointed out the network screens in the common areas, pronouncing the anthropomorphic interface's appellation "Deed ahh lus," rather than the customary "Dead ah lus." There would always be a large video screen "in your living room," she told them(him), and a terminal on a desk for surfing the Internet. An additional, suitably sized TV, for streamed entertainment, news, and communication, would be opposite the bed, a queen sized if he preferred. He preferred, he was quick to acknowledge.

Barnard attended to her stream of commentary, less for content than for its tonality, as the elevator carried them up again, past where they had been and past the seventh floor of the new addition, to the neatly arranged roof garden. The lightly breezy patio was shut in by low greenery on one side, black painted iron fencing on the others. Barnard took in the collection of sitters, amongst a scattering of perennials in mahogany boxes, without comment. He briefly surveyed the rubber-tipped walker and the silver spoked wheels hemming in an elderly pair in quiet conversation under sun-spattered, broad rimmed hats.

It was an unusually clear day. Barnard squinted and surveyed the sea's distant scintillation while Chen inquired about his interests. As throughout his visit, she endeavored to give no hint that she knew anything of Barnard or that she might be aware of what he sought or was likely to miss. She made no reference to what she had been told, that he had lived for decades less than a half-mile away and had raised two children there. She projected little awareness of his abbreviated naval career or of his subsequent academic position. She encouraged Barnard – imperceptibly, she imagined – to introduce these aspects of his life to her as if new information. How many times she must have made the same crafted presentation to a similarly divided audience.

*   *   *   *   *

Ten. Eleven. Twelve.
Miss Chen.
The cool flow from beneath her office door is, other than Barnard, the only thing that disturbs the air of the long hallway. He slows his pace to see if a glow reveals she is at her desk.
Not tonight.
Thirteen. Fourteen.
"Ah credit ration."
"Pfffhhhh," Barnard emits into the empty corridor.
Of that entire day, out of all that had passed among them during that round of introduction, Chen's peculiarly accented remark regarding the PT room's array of equipment has remained a mnemonic tease. "Ahk redit tation." Or perhaps she had put it as, "Ahk redet Hatian," he has often mused. Occasionally the polysyllabic figment will emerge from distant, indistinct babble. He has heard it in environmental noise, has extracted it even from the vascular hissing and pulse in his ears on sleepless nights. Her phrasing is an unimportant fragment of a distant day, one that nonetheless remains.
Picayune perseveration of a pointless particle of a piecework day. Proustian.
Need move along. Ice cream'll go soft.
Repetitious recollection is part of Barnard, as is over concern with details, with the obscure, the obscured. He has need of ponderables and has a large stock of them at the ready. How she articulated the phrase is of no particular significance. Barnard's frequent replay is ritual triggered by circumstance, just as is his enumeration of the eighteen doors (doubles counted as two) along this hallway, just as was his enumeration of the fourteen then thirteen, landing, fourteen then thirteen again steps before the landing at Three. Chen's phrase is a marker, a linchpin. It connects.
Bad idea, right, this suit.

The double doors of PT are wide open. Barnard pushes out a constricted puff of air, sensing that he had best moderate his pace, in addition to his racing thoughts.

Fifteen, sixteen. No one here.

Barnard briefly surveys the hulking array of equipment – shiny chrome or dull grey, oddly shaped, vague in the fading light. He prefers to be there when it is empty, when he can be free of conversation and observation, quietly trudging to nowhere as a prelude to a shower and, hopefully, less disturbed sleep. He takes pride in his physical appearance, in its echo of vigor. At five feet ten (once), his 169 pounds (wishful) seem to him to be well-placed, despite waistbands' evidence (shrinkage). Barnard judges that he is in better shape than his peers. This keeps him motivated. He has a nice suit that flatters. He can eat in the dining room or in his kitchen upstairs. He can be alone or be with Sallie. He can go to the OP Café, stroll along Lincoln, or visit Venice and its Boardwalk. He has his friends, Phil and Frank. During his workouts, he fantasizes vanquishing opponents. He imagines he would cope if accosted and would evoke surprise with his targeted quickness. He requires no guides, no helpers, and can double step the stairs. Were the need to arise, he could still carry three or four grocery bags at one time, as when Diane held the door for him and shook her head, anticipating the worst.

Dimming. She's dimming.

Staying active is important to Barnard. In summer, huge crowds often show up at the beach, which makes going there for his tensed taekwando less pleasant. He takes long walks on the Strand or Boardwalk instead. When winter weather does not cooperate, when it degrades beyond gray and chilly, he can use the ten pounders in his room for exercise. Far better than the walks or weights, however, is The Machine. While he is grateful to have it available, it is regrettably less inconvenient than he would prefer.

Firstly, he must go down a flight to the PT to use it. At home one much like it sat in a room off the kitchen, which previously had been used as their sitting area. Its French doors opened to the northwest, toward Santa Monica Pier, which they could not see, obviously. Still, as they sipped wine and talked on mild evenings, they could hear the vague tones of the merry-go-round, or the noise of the coaster's descent and the faint cries of its passengers. Barnard had replaced the rosewood serving cart with a large-screen TV and the lounge chairs with the expensive strider after Diane was gone. These extravagances, overly elaborate for his needs, were distracting and well worth the cost.

Secondly, while the Sunset Vistas PT room serves as his best alternative to outdoor exercise, he is unlikely to be alone there. Residents of Sunset Vistas are encouraged to be physically active, to use one of the treadmills regularly. "Go slow[sic]," they are admonished, "unless Doug Roach is nearby." As Barnard prefers not to use The Machine when the room is crowded, so is he especially reluctant when Doug is present. This provides the third negative, one often sufficient to turn the tide of Barnard's inclination. If he must, that is, if he finds Doug is in the PT, he will return to his apartment and sling the hand weights around for a half hour or so, marginally attending to a 'net feed.

With Doug on-site, it is easier to skip working up a sweat in the PT. How the man ever got to be an assistant escapes Barnard. Presumably, he keeps watch, maintains the equipment, and suggests the obvious to a range of residents. For this they have given him a partitioned off, almost-office with a desk. They should have engaged an outsider, a younger person, a professional, or at least someone in reasonably trim shape.   Doug a trainer? What could that oversized, disagreeable, and totally unpleasant boil of a person know about the needs of the elderly? Barnard has quite often ineffectually mused.

None of this animosity is new to him. More than an inch taller than Doug (in high school it was the opposite), he would love to expunge the pushing and grabbing, the teasing and stealing until Douglas (as it was then) had graduated and, at Long Beach State, been made aware of his own insignificance. Barnard even had given over his Best Summer Camper medal one fall, thereby being a double fool – for wearing it and for imagining he could appease.

"Hey! Barn Yard! What's that ya got on?" Barnard resurrects in stride. Douglas and his toady, Jack Harrick, had been prowling the halls at the start of the year. They had missed not having Barnard Cordner to pick on for an entire summer.

"Here. I don't want it," he had lied to their close stares, receiving a rough block into his open locker as reward. The coat hook had bruised his temple.

"Pheew! What's that stink? It's Barn! Yard!" Douglas had tossed over his shoulder with the medal as the pair laughed and merged into the bustle.

Inch over and ...

The memory is one of many that remain sharp for Barnard's typically nocturnal review.

Verisimilitude of sight or sound, similarity of circumstance or place can trigger replays of banalities as easily as of the significant. In addition, with emotion attached, one is taken beyond primary recollection. Writers have achieved fame by exploring this human trait of restoring portions of the past to present experience. Do animals relive in this way? Barnard often has wondered. Or do they simply process sensory input so as to survive, and thus reside in a continual present? Dogs do dream. Elephants do cry. It is a spectrum with humans at one extreme, he had long ago decided. The quantitative differences in the detail of our recall, not its core nature, are what separate us from other creatures.

Regardless, Doug is high in Barnard's stock list of accumulated disturbances.

Enough!

It is another and potentially final Sundae Nite. Passing the dim chamber redolent with Doug/Douglas, Barnard feels the impact of the sideways blow that had ejected the sector of pizza from his grasp one distant night in front of Martinelli's on Montana.

* * * * *

"Best Hand Thrown Pizza in Santa Monica," the window lettering promised. The nimble creators worked behind its clear glass, through which Barnard would watch, imagining what his choices of topping would be if he could order a whole pie. Standing with his re-warmed single slice, he would follow the rolling, stretching, then flinging up and stretching again of the mysterious pale, elastic dough, all the while anticipating the disaster that never came. Few have retained that method of making pizza. Stiff sheets boxed according to size is the current corporate rule. "Thick crust or thin – your choice" prepared in factories miles away; possibly even brought in by the planeload from dollars per day places that seem to have defined the world's destiny. The "Last Pizza Artist in Santa Monica" article in the old *Pico Post* had given them a boost in business, which, in part because of competition from the chain pie sellers but mainly due to Martinelli's compensatory cost cutting that diminished the appeal of his product, did not last. Barnard later adopted this for an example in his lectures and preserved it in his notes for The Book: The false enterprise economics of minimizing proximate cost, in place of maximizing long-term net gain.

Earlier during that particular day, more decades ago than he cares to count, Barnard, unwisely, had used the diminutive. "Doug," he had begun. The then Douglas did not care for that. His friends were thus empowered, no one else. It was late spring. As Barnard stood just to one side of the big "Best" painted on Martinelli's window, eating a slice of the pizza-man's creation, the tormentor strode up and inserted himself.

"What're ya starin' at, Barn Yard?" His back to the flying dough, he pushed hard on Barnard's shoulder. "Wan' a rabbit?" he then asked with a cruel smirk, skinny Harrick by his side, presuming Barnard's ignorance and not caring for coins or medals nor interested in the taste or smell of pizza.

Barnard suffered the palm up strike with no reaction beyond the involuntary. Empty hand slack at his side, he did not fight back. Neither did he ever again stop to watch the preparatory juggling of pizza dough. In addition, he did whatever was sufficient so as never to share space with Douglas Roach.

* * * * *

"Pfffffhhhh."

Barnard feels as much as hears his constricted exhale.

Enough. Scurfy-faced shit.

It is not true what they say about bad memories. Like the fabled pea, they can disturb no matter how numerous the intervening layers. Barnard perhaps prefers it that way, viz., subconsciously and paradoxically choosing to have the details of unpleasant pasts remain accessible. At least this provides ample opportunity to remodel them, to alter what follows from the images. Still, his complexion takes on a darker shade as he strides past the PT. He never would have agreed to Sunset Vistas had he known Doug would be there.

An aide!?! Prick.

Took karate too late. Too bad. Fist strike and takedown would have ... Right.

Reliving the past, with the insertion of different responses leading to different outcomes, is reverse practice. It precludes, one would hope, repeating the mistake. True, the past is unalterable. Still, something much the same may come again. Rehearsal does provide an opportunity for fabricating the means with which to blunt similar blows, to transmute them or, by clever stratagem of word or action, to avoid them. The true gain comes, however, from making the present tolerable. At the core, all that perseveration changes is how you feel, what you feel. This is reality. Even if elaborating a sharper, more fitting response does not erase or change the event, it is a valid exercise. At a minimum it modulates the affect of that which is immutable.

Enough!
Cheeks'll be flushed. Give wrong idea.

Barnard lightly strokes his cheeks. He pushes ancient images aside and, aware of the heat on his face, imagining the coloration thus imposed, navigates the final few steps down the hall. He thinks about The Machine, the pair of ten pounders, and taekwando, about the overpowering sea and yielding sand.

"Forget the bastard, already," he utters aloud, knowing that he is alone. "Tomorrow, a full hour. Before dinner."

Chen. A Credit Nation?
"Pfffhhhhhh."

Barnard hears murmuring ahead. He rotates his head, working his suddenly scratchy shirt collar against his neck. His breathing is noticeably deeper. He is either walking too fast or musing too intensely. Or, it being a weekend night with the AC purposefully off, it might be because of the slightly humid air.

Why? Cost. Heat rises, right? Puzzle solved.
Seventeen. Eighteen.
Show time! Time to –

"Shit. God damn it," Barnard hisses.

Doug is barely inside the entry ahead, talking to – if memory serves – a nephew, who exhibits the same bad skin. Barnard intuits that the latter suffers from the same personality as well.

"Damn it!" he repeats.

He hesitates, takes a few short steps then tacks down the side hall to the men's lav. He has already gone, so it is a feeble stream, slow to stop. He should put a few folds of tissue over the tip after, he decides.

Don't want to spot. Get me one day. Everyone gets. No, not everyone.

The stolen pun evokes a small grin as he rezips. He washes his hands carefully. It has been years since he has had to suffer the indignity of a digital exam. His grin widens.

Every cloud has its –

"Damn," he barks at the access panel above the urinal that continues to flush. "God DAMN it! You son of a," he begins in a resigned, hoarse rasp. Hot blood again infuses his face.

Cheap, porcelain, chrome-topped piece of shit!

Barnard rotates his head left then right in mute signage of "Why the hell don't they fix it?" A second hard pulse of annoyance tenses his jaw. He jams the lever down, then up, down, up. This helps. He wishes it would break off; that would finally justify a plumber. Down, up. With this, the flow of water ceases. He stands quietly for a moment, inspecting his hands, then he washes them again. He presses his palms, damp after using the coarse paper towel, against his hot cheeks. This feels cool, soothing.

If only color would not come so easily to his cheeks. It reveals him to others, often revealing more than of which he is aware and always more than he prefers. He would make a terrible poker player, a poor negotiator. He tries to force his mind forward, to his upcoming sundae and whether they will have new flavors tonight. He fails.

"Barn Yard."

Nasty prick.

"Forget it, already. Decades ago," he hears. The reverberation of his own full voice startles him. He looks to the side then over his shoulder, grateful to verify his solitude.

Exercise. Tomorrow, at beach. Or on Machine. Right. Before dinner.

"Uh credit nation." *That's it. Right? Prescient? Irony? No, not Chen. Not her way.*

"Barn Yard." *Ignorant son of a –*

"Stop it!"

The overtones of his exclamation resonate in the narrow channel leading from the lav. Anger once could be a useful tool. No longer. For Barnard, it is a pointless relic, yet one he cannot discard. It still requires release, even if misdirected and of a trivial nature. He knows perfectly well, were he to be asked, that of the trio of glistening porcelain urinals, the one in the middle, the one he used, is uniquely prone to malfunction.

Trial and release. Test and verification. Time and time again; for that there is the need.

CHAPTER 3

PAYBACK

Standing at the threshold of the activities room's double doors, Barnard looks over those in attendance. For the Sunset Vistas ladies there are few Sunset Vistas men, he notes. Despite the recent changes in lifestyle, politics, and workplace, the fact remains that nature puts men in the minority near the end.

Genetics. Stress. Die off earlier. Modern Women will do likewise. That's coming. Right. Conflicts and obstacles; obligations and demands; money and careers. It'll even out. No big mystery. It'll even out.

Barnard struggles to restrain his self-directed dictations. He is acutely aware that he must discourage them before they bloom into internal rhetoric. Easing in, taking short steps, Barnard masks an unintended veer to port with a quick pull on his jacket cuff. At least Doug and his nephew are no longer in evidence.

The ladies seldom run out of topics on Sundae Nites. There is always a complaint or health issue to (re)consider, received images to share (again), old gossip, or a new arrival. Many purchased their apartments because there was little else for them to do. Less practiced than the men at living on their own, that task was judged to be fraught with risk. Others are in residence by virtue of convenience facilitated by formerly sensible financial planning. Some are relatively young, new to Social Security and pleasantly alert, ready for a visit. Being open has grown easier for Barnard. He no longer is startled by a glimpse of Diane's profile, rarely sees her taut hair from behind, or, with no specific target in sight, ever detects her scent amongst those present. Gone for over a decade, she is finally absent. She, the boats they sailed, the home they shared, the two children they raised with mixed success, are memories that have grown faint, like the photos of a young Grandma Jane-Jane that are stored – cracking and spare of color or detail – in a box high in his narrow bookcase. Why don't all memories fade like that? he has often posed while staring into formless darkness. Some have, no doubt. Untested, one never can be sure.

The gentlemen come to the get-togethers typically to continue discussions, to win arguments that began elsewhere. Living alone, being bombarded by screen images, tend to make everything glibly well-defined, easily understood, and eventually stale. The men therefore drift into repetitive, never resolved, most often factually irresolvable debates focusing on politics, The Market, diminished portfolios, or sequestered funds. Facing down a few gaps in their points of view innerves them. To have exposed some others' misunderstanding of the world's current state, potentially to elicit a few murmurs of agreement, is enlivening. The way male minds and muscles are coordinated for action must account for this. Barnard prefers to stay out of their scattershot discussions. He knows too much, feels too much, and is far too serious.

Since the implicitly coupled Black Weeks of the Big Crash, there is no escaping the dramatic changes. In this regard, the ladies, reveal their increasing insecurities easier than do the men. They tend to dwell on how they have been personally affected and are less inclined toward argumentative posturing or socio-political fault finding. Barnard sympathizes with their typical unconcern for establishing proximate cause, especially on nights like this. Sundae Nite is a time to put aside current worries, to replay old rituals and revive pleasant memories. It is an opportunity to don something other than age-specific leisure wear, to retrieve something from a crowded closet, a favorite outfit that might have been worn to the office or when out for the day.

Barnard buttons his jacket and sniffs. There is something familiar in the air.

* * * * *

Climate made Southern California a popular final destination for seniors. Once governmental programs in support of expansion of retirement facilities were created, the easily obtained, federally guaranteed low interest loans were hard to resist. This was not to be a simple-minded sector bubble, a soon-to-crash folly like the poorly sited strip shops and minimal qualification single family home fiascoes of the noughties. Real estate had been in recovery since the early teens. Retirees had their IRA and K accounts. The programs' prospects were solid, notwithstanding the plaints of harpies who could not forget relatively recent events.

"This is California!"

"They'll need to live somewhere!"

"There's no other way!"

At least this was the plan.

Many of the larger complexes in the area were transformed. The enhanced Eastern Star was renamed Sunset Vistas. Well-located, between Santa Monica and Venice, it had been first class and expensive. It was thus a prime candidate for conversion to purchased apartments. Barnard's mother, Frances Graese Cordner, a longtime resident, was spared that noisy, dusty inconvenience. A few days after her funeral, Barnard gathered up her few remaining mementoes, took a last look around the third floor visitors' lounge, soon to be endowed in her family's honor, and anticipated that he would never see the inside of that building again.

Once its additional floors were complete, Sunset Vistas was a prime retirement facility. Golden-agers benefited from an annuity allocation of the savings that would otherwise be exposed to market risk. Equity terms were generous. When a resident's unit was vacated and resold, a portion of the original investment went to his or her estate. Until that departure, each was safely ensconced in an apartment purchased with firm assurance that meals, activities, and medical oversight would be provided. The reconfigured space, in particular the three uppermost floors, enabled Sunset Vistas also to provide graduated enhanced care. This implicit internal migration provided another mechanism for freeing up the lower, reconfigured larger units for resale.

Thanks to gifts from his aunt, Laura Graese, and the Graese Foundation, an endowment for in-house amusements had been established at Eastern Star and carried forward in favor of the Sunset Vistas' residents. In addition, the events space had been provided with a fine collection of quality furniture, similar to the pieces that Barnard remembers seeing in Grandfather Adolph's and Grandma Jane-Jane's dining room.

Unfortunately, the European style tables, with thick tops and heavy legs of dense, straight grained wood, are mostly gone. Only one, this evening covered with a plain cloth, remains. The others, never properly polished and frequently wiped with damp toweling, aged prematurely. Careless hands spilled, dented, and scratched. Many, Barnard suspects, were borrowed – relocated, as he might injudiciously say. The few matching heavy chairs that remain have been shifted downstairs to the common room and entry area. Stained and frayed, especially along their broad arms, they are safe. The prospect of expensive cosmetic repair makes them poorer candidates for surreptitious appropriation. While they serve as hints of prior substantial support, a critical eye would judge them to be unsuited to a facility catering to seniors with stiff joints and weak arms.

Since Barnard purchased his apartment, the activities room has been inexorably transformed from elegantly ill-suited to communally functional. Its purpose now is served by a mix of wood composite tables with metal edges, plain rectangles covered in faux wood grain embossed plastic. Scattered about are squares of light blue on skinny metal legs, the sort of tables generally found in institutional cafeterias. Chairs, most of them with lightly padded seats and backs of an age resistant shade of brown vinyl, are indiscriminately placed. The contoured plastic stackables, recent additions in an even less attractive shade of brown, are mainly for the benefit of visitors, for an increasingly unlikely overflow crowd. The changes parallel Barnard's recollection of his own early circumstances: fortunate, as when he was a child, giving way to the less pleasantly modest, as after the premature loss of his father.

The institutional aura has not dampened appreciation of Sundae Nite. The event endures. It offers a welcome respite, an opportunity to be immersed in a convivial setting consistent with the limitations of age. Few fail to perceive that it is more sensible to reside at Sunset Vistas than to be burdened with the upkeep of an independent home or apartment. Unfortunately, what the residents had attempted to control, to plan for, is increasingly for others to control and to plan.

* * * * *

Pizza slice? Not likely.

The odor Barnard detects is a confusing blend. He is too human to disentangle the mix and yields to a less complicated present. His proximate target is Sallie; he wants her to see that he has arrived. She is at the far side of the room. Erect, slender, with still naturally dark hair restrained by a barrette, she is speaking with May, her ever present friend. Sallie moved to Sunset Vistas the year before Barnard arrived. Her decision, several months after the loss of her husband, Gerald, was easily made: She was childless and had no extant family, neither of her own nor Gerald's; being newly alone was heavy, monotonic. She had looked forward to the simplicity of a compact apartment with plenty of people about. Through unpredictable intersection of trajectories, Sallie had found her former best friend already settled in at Sunset Vistas. In a matter of days they again were constant companions. As their yearbook had prophesied: "Sallie Bass and May Westerbrook - Most likely to remain friends." There had been a long gap.

They make a superficially incongruous pair. Sallie's face is a creamy oval, presenting a strong chin and prominent cheeks. May's is ruddy and round, beneath short white hair and set upon a stocky frame. In high school May had been wild, an undiscriminating hellion, Sallie more reserved, selective. Sallie has described for Barnard how popular May was, that she had been petite and forward. As hard as it was for him to visualize the former attribution, he has always refrained from making any relevant remark.

The formerly inseparable pair had remained close during their single years after graduation – their intimacy unburdened by contemporary implications – then had drifted apart, slipping into the confines of their own structured worlds. Although they honored each other's birthdays with cards and calls, they communicated only occasionally and met rarely, typically for lunch.

Barnard once again imagines that they probably exited the sixties with not much to show for it save early marriages and suburban ennui. He smiles upon Sallie in the distance.

* * * * *

Retirement collectives are replete with incipiently bored, prior strangers who only tentatively learn each other's stories. For May and Sallie, linked by a rich set of distant memories, to again have each other for company is a mutual benefit neither would have dared hope for. Sharing the details of their intervening decades occupied hours of earnest conversation. For a change of scene, the chattering duo would take the bus up Lincoln to wander among the shops on the Third Street Promenade and to enjoy the novelty of cheesecake or flan with overly roasted coffee. The two ladies found it pleasant to simply chat at a sidewalk café, the warmth of the sun moderated by the light shore breeze. Recently, not totally attributable to finances, their outings occur less often. They are feeling their age.

And there is Barnard.

When first he met Sallie, he had sensed a mutual attraction, which, being authentic, soon grew into fondness. While they do share most evenings, Barnard will not sleep at Sallie's apartment. He finds her rooms overly feminine. Instead, Sallie goes to Barnard's apartment to sip wine, to talk, to watch episodes of weekly programs, specials, occasional trash, ersatz news, and old movies. When she stays overnight, it is for uncomplicated companionship.

Even before Diane died, Barnard had gradually accommodated to being alone at night, in respect of her fitful discomfort during those last months. Her clinical course, as both perceived despite verbal denial, was guided by palliation not efficacy. As the burden of their foreshortened allotted time became heavier, neither benefited from physical closeness during long, uncomfortable nights.

Now, the pleasant warmth of Sallie's leg against his is most welcome. He does not want to be alone again, even though there are those nights when he would have preferred to keep the window fully open to accept an onshore westerly and the faint sound of casually rhythmic surf. For him to have found her must be attributed to ignorant accident. As he had not been seeking a companion, neither was there any particular reason for Sallie to have elected to come to Sunset Vistas. She could have chosen any of the other local facilities or even moved out of state since, with Gerald gone, there was nothing to bind her to Santa Monica. Additionally fortuitous was May being in residence. The providential tripartite arrangement has allowed Barnard and Sallie's relationship to develop without becoming possessive and thereby short lived. As much as she has grown fond of Barnard, as much as they are fond of each other, Sallie will never give up May. It is satisfactory, for all three, that Barnard has her evenings and May has her days.

Some would suggest an underlying benevolence, would even invoke some overarching plan. They, Martin Stoole and others who hold this to be always the case, would love to discern that Barnard, even for the briefest, tiniest, most infinitesimal of moments, has favored such a theistic reality. Because of his thoroughly secular bias, they see him as a challenge to their formalized faith. Few, however, could be less susceptible than he. In contrast, most, certainly most of the residents of Sunset Vistas, cannot get enough of The Almighty. God, of a particularly specific and richly defined sort, has become The Thing – the HBFF (Heavenly Best Friend Forever) of tweets and texts. Retreat to such presumably firmer ground invariably occurs in difficult times, when the otherwise familiar can be uncomfortably disjointed and unreliable. "The Devil made me do it" in the good days is less cogent than is "The Lord's Will" in the bad. Thus, the authoritarian, political Theocrats have put to "good" use the economic disarray and its accompanying pressures. These facilitate, even encourage "Trust in God's Plan," as Senator Whatever-his-name-is keeps urging.

From Barnard's perspective, the pervasive theocratic tenor demonstrates "profound confidence propelled by a paucity of conception." He enjoys that construction and has repeated it often, as well as having jotted it down on a notecard for inclusion somewhere in The Book. Nevertheless, they, the domestic fundamentalists, are winning. Barnard has been forced to acknowledge that the constant theo-political harping – on God, His Plan, and What He Means For Us To Know – is effective. His personal inclination is to debunk the whole of it. He is frankly annoyed when he senses that anti-secular social vector, that dogmatic nudging. The current of popular governance is strong enough. Seeding it with the winds of divine authority over a broad fetch of popular ignorance has yielded an overpowering wave, as the former sailor might say. He resists it as best he is able.

His own move to Sunset Vistas came a few years after Kyle and his family moved eastward, to Texas. Barnard more fully appreciated the onerous aspects of owning a house once he was relieved of them. This made it easier for him to adapt to the new lifestyle, one clearly wise to accept before it was mandated by physical or mental limitations. There were excursions, to Newport and Catalina, for example, and up the coast to Santa Barbara and Solvang. A pre-opening tour of the re-remodeled Getty was notable, since Barnard frequently regretted the sale of his collection of German art prints. He chose to decline handicraft workshops and community involvement, those proffered activities that seemed to yield less than they required. In partial compensation, there were cooking classes, lectures, and special video presentations in the media room. Most of these have been discontinued, but there still are Sundae Nites and the neighborhood OP Café, being close to the Venice Boardwalk, the Santa Monica Promenade, and the beaches. There is still the faint pulse of the ocean in the distance, which soothes him on difficult nights. In sum, his apartment has been a pleasant stepping stone between independence and eventual managed care, if he lived that long. And there is Sallie, with whom he shares pleasant moments. Unfortunately, recent financial events have made their present less comfortable, their future uncertain.

*****

Shifting his glance from Sallie to the others in the room, Barnard simultaneously attempts to prefigure some favorite entrée, a dish that could have been the proper prelude to his anticipated assemblage of ice cream, syrup, and nuts. Three-day sauerbraten? Jaegerschnitzel? Their reification eludes him. The reality of today's mid-afternoon buffet will have to suffice. He redirects his attention to the room's faint odors and murmurings, to the high ceiling so different from the rest of Sunset Vistas. The floor is smooth, polished, with a faint random pattern of industrial black, gray, and green. Bright, narrow-beamed accent lights set in among the floods glint off the metal surrounds of the chairs and refract through the sweaty water glasses on the table, projecting bright sprinkles at his feet. Barnard is struck by the impression of a calm sea under high sun. He looks across the space between himself and the dessert table, at the beckoning goal that it represents. Someone bumps into it, hard, jostling the iced water glasses. Agitated multiple reflections dance about on the floor, hinting of head wind and chop. It reawakens also an even dimmer past.

Judith.

Evoked by the skittering bits of light is the sparkling ball high over the old Y's dance floor. The first flowers for each, which he had tentatively bought from a mid-block florist, come to mind: his stabbed through the buttonhole of his lapel; hers dutifully pinned, by an unsmiling mother, to the strap over her left shoulder. They were strong, pale shoulders ripe with a light, flowery scent. He can almost feel the beat of the Hokey Pokey, can almost hear the slow music for dancing close. Barnard's fingers extend, to touch again the stiff material at her waist. Arms out then clasped in, Turn, turn, and step, her forehead against his chin. Then closer, bending and blending as one, immune to the crowd, warmed by urges that were still new. Turn, step, step, turn and dip, his lips against the bouquet of her hair. So close. Pizza in the dark by the reservoir after. Alone, unseen, and so close. Hungry to know.

He elevates his cheeks as he strains to reinforce his present, this Friday evening, this Sundae Nite. He scans the room for Sallie, this time without success.

"Pfffhhhhhh," he sighs.

* * * * *

The activities room of Sunset Vistas was formerly well outfitted, the facility fully staffed. In the Graese Foundation it had a firm source of support. The residence had been burdened neither by an over abundance of selective authority nor governmental financial assistance, the latter often serving to make the former acceptable. Thanks to the Foundation's generosity, periodic early evening buffets, even a cash bar (two drink limit), were provided. These were notable events, comparable to a night out. Occasionally there were real musicians. Hence the name: Big Band Night, every other Saturday. Now it is second Friday Sundae Nite, with streaming music from speakers in the ceiling. The bar is gone. The stated preference for no alcohol in the apartments is an unwelcome restriction, one which Barnard has consistently ignored. Recently, he finds it less onerous but for the wrong reason – expense. In any event, it serves to illustrate the current societal reality: Couple the power of financial responsibility to a sense of a higher moral purpose, then control over personal behavior is easy to justify. Add a dollop of epidemiological ambiguity and there you have it: Opportunistic micromanagement masquerading as concern.

Theocrats have been working on attaining this leverage for decades, for centuries some would argue. At last they have attained it. One growth of traditionalists has been chopped down and another has taken its place, like weeds in a field. Barnard has often gloomily explored the eventual consequences, as revealed by history. A retired academic economist, one who prefers to wear the mantle of a liberal humanist, the authoritative impositions offend him on basic as well as practical grounds. How ironic it is, he has argued, that Theocrats, who for so long and with so much fostered rancor decried governmental overreach so as to claw themselves to its head, have established their own style of hubristic imposition. It took a few false starts and several practice runs, but they knew precisely what they were doing and when to do it. Those seeking power thrive on chaos. This has come, in great measure. They are not letting the opportunity go to waste. People are quick to accept limitations when they come with benefits, even when those benefits are little more than partial cancellation of woes recently imposed.

Barnard still retains his firm intention to elaborate upon these ideas in The Book, his long-standing project of a behavioral economics text.

Early in his retirement he had made good use of the nearby Santa Monica College library to work on it. It was not that he had to. The Internet sufficed as a resource. He could remain in slippers the entire day if he so chose. Nonetheless, the library was familiar. He was at home there and could more easily concentrate. For a while it did provide the occasional isolation that facilitated the focused effort that the project required: organizing his notes, formulating, setting down. Meeting Sallie and feeling a reciprocated natural attraction gradually displaced these intentions. The continuation of academic pursuits was probably not a realistic plan from the start, he eventually concluded. However, because of the recent financial disarray and the ways in which it is directly impacting their lives, a sense of urgency has returned. The project has taken on a renewed importance for Barnard. He is again inspired to compose segments of focused text.

His project's origins predate his formal training in economics, actually. When he was an undergraduate physics major, a Pieter Bruegel painting of peasant life, entitled *Wedding Dance*, was a prominent feature in the lounge of Barnard's fraternity at U Cal Santa Barbara. It was a fine professional copy and his pledge task was to clean its frame. As he worked a dampened toothbrush into the details of the deeply carved wood, he was drawn to the formal realism of the work. There were kissing couples and a cutpurse or two. The dress of the participants, the quaintness of the cod pieces, coifs, and aprons, the earthy coarseness of their dance all seemed to match the rough nature of the space in which they cavorted. However, the vagueness of their smiles belied the depicted frolicking. Several of the revelers appeared to be looking upward, as if at the viewer, or at someone or something higher.

Off to one side stood a forlorn peasant at her task, holding a heavy basket on extended arms. Most of all, it had intrigued Barnard that at the periphery of the gathering, far into the distance, an isolated figure in a broad, black hat stood facing away, with hands clasped behind as if in deep thought.

The Wedding Dance - Pieter Bruegel c. 1566

Detroit Institute of Arts, USA (City of Detroit Purchase)/Bridgeman Images

Years later, as Professor Cordner would prepare his lectures on the societal ramifications of academic economic principles, that painting would come to mind. He would recall the depiction of patient, constrained celebration. It resonated with his increasing dread of an Orwellian progression to an age of universal, therefore planned satisfaction. Bruegel, he imagined, could have been depicting exactly this concern, his dismay at his own era's acceptance of imposition masquerading as pleasure. Clearly, the painting was not created for any of the carefully represented, less than jovial peasants. For what or for whom was that lone man in the broad, black hat searching? Barnard had wondered. The diminutive figure was obviously significant. The entire composition focused on it – the converging lines of attendees on either side, the apex of the triangle of dancers in the foreground. Could it represent the artist, be a hint that he understood the nature of, and therefore rejected, the merriment he was so adroitly depicting? Barnard was sympathetic to this view. Observing the rapid growth of consumerism and of marginal satisfaction masquerading as total, he suspected that being reduced to accepting an orderly, useful, and predetermined level of enjoyment could be the inevitable end point of any consumer oriented socioeconomic system, the paradoxical result of domination over rather than by natural forces.

Government, insisting on stability and protection of its assets, has indeed taken control of what was once solely for Barnard, for each of them at Sunset Vistas, to manage. To a behaviorally oriented economist, they are on "the cusp of another wave of feudalism, a regression to the world of allowed pleasure as seen by Bruegel, a return to serfdom." Barnard considered that snatch of prose to be worth saving for eventual inclusion in The Book. He has it, also, neatly inscribed upon a three by five card in the file box along with the many others.

* * * * *

Barnard takes in this evening's assembly, the similarities of posture and manner modulated by age. Yes, they are getting older. Yes, their funds are tied up, under external management for the most part. Yes, he is one of the misfortunate middle, neither so poor as to have lacked aspirations nor so rich as to have escaped concerns. He had planned. After a few career missteps he had laid out a future and prepared for it with focused deferral. He had played the farmer husbanding the successful harvests for seed and the lean years. As is often the case, circumstance intervened to subvert his planning. "Intruded" may be the better word, but it is his word, not theirs. He shrugs, affirming that there is no point in letting distant remembrances or current worries diminish this Sundae Nite. It is enough that he fears for the continuance of the event. He should put this all aside and simply enjoy what is here.

What a shame, he muses as he inspects tonight's attendees, that he cannot recall anyone with whom he might, years ago, have shared a superficial dalliance and, of course, suffered a subsequent pang of guilt. The lack is real, not faulty memory. Those he had met at distant academic functions would not be expected to be part of his current world. But there were local admin aides and ambitious wives of colleagues who had noticed his surreptitious glances. Any of these would be of the right age to be here. Still, there had been no one. Not once. Never. Diane had been all he needed. He had not even succumbed to the coeds – in tight jeans or skimpy skirts, pouting for a beneficial adjustment of grade – for whom he might have shut his office door, as several pleasantly hinted.

Looking about aimlessly, Barnard feels the faint regret of having missed out on a memorable adventure or two. He could use her, if there were such a she, to tease Sallie as well as to elevate himself. No one here or in memory can truly claim that role. Except for Condi Carter to some extent, perhaps. Yes, Clifton Carter's wife does come close and she is a resident of Sunset Vistas. He looks for her through the gaps in the standing attendees.

Where?
Silly. Never would have. We ...
She ... I ...

The memories flicker and fade. He will revisit them later. He shrugs again, this time shaking his head to solidify an end to unwise fantasies. There is ice cream to attend to. The array is nearby, as are Sallie and May, whom he does spot again. He will join them after acquiring his ice cream and his slice of cake. In respect of the results of prior poor judgment, he then decides a sundae shall suffice. Cake demands coffee and their so-called decaf often earns him an interminable night of fitful sleep. Besides, he enjoyed a large helping of cherry cake a few hours ago. Neither will he be heavy-handed with the chocolate syrup. Barnard no longer anticipates the traditional solid eight. He endeavors to gracefully accept unbidden interruptions from within or without. Often there is the three a.m. stiffie and, virtually without fail, there are the careful treks in the blue-green glow to the bathroom. There are the blows off nearby Santa Monica

Bay and his occasional stumbling to fully close a whistling window. There are the barely grasped noises in the hallway, the assignable bumps and squeaks of rubber wheels marring a wall. It makes no sense to deliberately chance opening the floodgates of rumination via inadvertent caffeine.

The decaf memo "To: Ms. J. Chen ..." for example, was intended to firmly convey his complaint regarding mislabeled coffee urns. Awake and fitful, he had composed several paragraphs detailing the laxity of careless staff "... in this instance. Not 'care-less,' mind you; they do try." He had included this disjunctive, partially contradictory phrase in each rewrite so as not to appear unduly churlish. The exercise was pointless. Nothing could be expected to change, with or without polite temperance, with or without actual transmittal.

Barnard's late night ramblings represent more than unrealized text, more than stagnant missives studiously edited on the blank page of the dim wall opposite. There are always arguments to reform, better ways to have spoken or to have written, world events to set right or, at least, to expose for the less attentive. Barnard's rehearsals and post hoc critiques are part of him. But, he would insist, they do not define him. Neither is he troubled by being often hesitant in his speech, given to pausing and not finishing sentences. This is a mannerism enhanced by being ambivalent, by being unsure of whether he wants to express himself indelibly. When he envisions he is giving a lecture, presenting a paper at a conference, or even a local and informal invited talk, the problem is reduced, his hesitation disappears, analogous to the way a stutterer can often sing without defect. The artifice can be quite effective, if he manages to avoid affectation. However, it is only in fiction and myth that such monadic attributes are truly definitive. In real life people are far more complicated.

Being a late reactor ranks far higher on Barnard's list of intransigent deficiencies than does the hesitancy itself. That he rarely comes up with the quick, on point, and sharply apt response he takes to be a notable flaw. Foresight does not seem to relieve it. Clever rejoinders do come but, due to their latency, rarely satisfy. Therein lies another reason why he perseverates, why he indulges delayed, ineffective reviews of what should have transpired, what other ways he might have acted, what additional words he could have uttered. The precise argument, the succinct phrase and reinforcing gesture are revealed to him well after the fact, usually in the midst of staring down a sleepless night. Serial, relentless recompositions of recent or distant past can be as much cause as result of his occasional bouts of insomnia; there is no practical difference. Such replays do reward him with incipient corrections. Sadly, the luster of any, often brilliant burnish is lost by morning, like gleaming naval brass gone dull. Even if partially retained, it is an impotent reconstruction outside of context, is beyond what the linear progression of a singular reality permits. While each is invariably an improvement, the past remains. To placate the rustle and pulse of attendant adrenergic excitation, in one ear then the other as he alternates position against a firm, cool pillow, he must force himself to think of other things, to dwell on an impersonal observation, a recent video, or an old film.

Shadows of his nighttime musings also, of late, can encroach upon Barnard's days. It takes so little. A snatch of tune, even an instant of total silence or of the wind annoying the shade may do it. An unexpected odor – of wet metal or of scent upon flesh, for example, or of seaweed or flower – can elicit pregnant reflection. A splotch of light or an ill-defined shape can unleash old realities. But at night, when sleep does not come and distractions fail, the past can be relentless, can be laden with affect and profoundly real. The fecund umbrae of decisions poorly made or left undone are often cast over him, as had been the hot crimson of forced smiles. This trait is his legacy from a pliant mother, a cold Aunt, and a rigid grandfather, a molding untempered by a father who was too early dead.

Barnard breaches. He calms the roiling wash of old unpleasantries and floats upon the present. He scans the sweets awaiting his selection, accepting that, even when unaccompanied, chocolate cake is fraught with risk, a potential error with only himself to blame. His face feels hot as the anticipatory rejection arises: "No cake," he whispers. The corollary is intuitively obvious:

And NO coffee!

He looks over the familiar attendees, most infrequently approached, in a single sweep from left to right. There are exceptions, of course, since Barnard has not succumbed to true isolation. He takes notice, for example, of Thomas Schmidt at his spot by the entry. Tommy always wants to be, must be ready for that quick dash to the men's room. In control but wary, his puffy countenance will freeze and he will stare blankly into space, as if reaching for a far off sound, before shuffling quickly to an exit. Next to him is Al Silver – slim, neatly attired, passably handsome with a formal and cold personality. He hones his knife edge with the tine of a fork, Barnard has noted. Fastidious and neat, he always wipes both clean with his napkin before eating, which he nevertheless performs with gusto and impolite noises. In Barnard's view, the man has an unpleasant, dramatic, almost arrogant bent. Happily, he has given up trying to draw Barnard into discussions of financials and the future. This night he looms over the seated Schmidt, gesturing at the array of treats. Barnard finds the juxtaposition comical.

Still here?

His grin hardens as he continues to explore the room. He spots Frank sitting alone. Frank, the once boastful classmate with the big chest. A Three Letter Man. Then. The framed and yellowing newspaper photo still leans against the rear wall, lowest left, of their high school's trophy case – one man covered with eight. "Kurchak Carries All to Victory," is the headline, over a full column write-up on page C-3, lauding his last minute November touchdown. Pallid and physically slower, his damaged knee, from the cutback block (the current term) on a different, bitterly cold Thanksgiving Day, never fully healed. It was of no practical significance to his life in the wholesale notions trade and is infinitely less significant now.

The ruminative professor eyes the unimposing figure with the scattered hair. No intolerant swagger, now; no provocative smiles at giggling sophomores or arrogant jostles in the stairwell, now. Those have faded to insignificance. Frank and Barnard have aged *in situ*. The pungent burden of envy discounted by academic prowess, which seasoned their mostly insignificant early common past, has long since sublimated to casual friendship, made closer at Sunset Vistas more or less by default. Several times a year their paths would cross. At Santa Monica Pier, for example, Kyle and Katie would explore about while Frank joined Barnard and Diane to take in the sunset view from the fishing porch. There they would squint out at sailboats and the indefatigable striving of nearby surfers. They had enjoyed the food vendors, with their stalls open to the sea, and the upper level of the fish restaurant, with its constricted view and harried staff. They would linger in the carousel building, with the steady drones of organ and straining motor providing counterpoint to the rumble and hiss of the nearby surf.

Barnard is unable to recall much of Frank's wife. They had never socialized as couples. It was somehow sad that he, Frank, long married to the same pleasant but rarely seen woman, had no children to treat or to brag about. For the two gentlemen, each with a plethora of work-related constraints, it was mainly an occasional coffee, accompanied by a roll and small talk, when there was a break in their respective busy schedules. Their tenuous communions would take place without firm intention after a convention, trade show, or sales meeting for one, or after a professional conference, academic site visit, or distant lecture for the other.

Frank had developed a lucrative business jobbing the cheap impulse items that are put out near checkout counters and on holidays. It was mainly travel and details, smiling and sweating in those early years – the sweat of dollar per day Chinese, Barnard had on several occasions observed unkindly. Frank would counter that higher education, when not put to overtly practical use, was merely to be tolerated. Later, when Barnard – "Barn the Brain" in high school – was Professor and Chairman of Economics at SMC, he proposed that Barnard arrange for him to mentor a few international business students. No doubt in part a reflection of an unassuaged sense of inferior status, Frank felt he possessed relevant experience, not just dry facts, that would be helpful to them. Given Frank's coincident financial support for trade related special events and invited lecturers at the school, Barnard could hardly refuse. He made the appropriate initial introductions, thereafter letting one of the junior faculty handle the details. Barnard discreetly verified that the students found Frank's tutelage to be beneficial and, there being no complaints, left well enough alone. The semiformal arrangement yielded more frequent opportunities for them to have lunch and share the events of their lives. Barnard was sensitive enough to show only superficial curiosity regarding his friend's eventual regulatory difficulties.

When they found themselves sharing full retirement at Sunset Vistas, their friendship became more substantive. Frank has adapted, but is less inclined than most to enjoy the media-based amusements. He, much like Barnard, has expressed his suspicions about the underlying methodology. For example, only after numerous reminders did he agree to complete his Biograph. To be fair, he had little incentive. His wife died long ago and his sole surviving relative is a younger brother, also childless, located abroad. He has no descendants eager to explore their genealogical links. Also in his defense, explicitly constructing a Biograph has been rendered redundant by VieGie. The latter, the remote visitation system that warehouses and integrates personalized voice, text, and imagery, is a virtually automatic collator of the facilitated interpersonal interactions well as the data exhaust generated by messaging, shopping, social interactions, and even simply being entertained via the Internet.

Moving closer, Barnard catches a whiff of Frank's familiar musky aftershave. He motions a greeting. Before deciding whether to speak with him, he looks down to discern if Frank is wearing his distinctive, complicated socks. He is curious, also, as to whether his friend will evince the lapse of secure contact that Barnard has recently come to suspect, if there is any change that hints of the onset of a decline, which, if gradual enough, can go unappreciated by the afflicted. Innocuous conversation should suffice to detect if Frank is on his way to ZZZ, or to Relocation.

Heart attack and poof!
Or AD and Zombie Zone.
Death preferable.

Zombie Zone Zeven, Barnard's flippant impudent appellation, is that uppermost, seventh floor of Sunset Vistas, the most intensely staffed section. They are not coy about its designation: the Alzheimer's Unit. Barnard, in contrast, prefers not to utter the eponymous phrase. In the continuum of managed care envisaged for the expanded Sunset Vistas, Zeven is its apex, its concluding step. After his introductory tour of Sunset Vistas, Barnard has had no intention of revisiting that upper residential wing.

* * * * *

Barnard, with son Kyle and his wife, stood behind Julie Chen in the elevator as they left the roof patio and descended to the floor below. Facing away, her explanatory monologue was interrupted by a chime followed by a digitally enunciated "Seven." Her pause was evident and, if any of her audience had taken a moment to reflect upon it, revealing. She stepped out, briskly leading them through an open area with a staff station at the near wall. The latter, consisting of high counters enclosing a desk, was decorated with a telephone handset, several monitor screens, an open laptop, a misaligned pile of folders – probably charts – several trays, and a scattering of accouterments of the medical trade. The few indistinct voices were no match for Julie Chen's. Kyle focused his attention upon her, doing his best to appear attentive. Periodically he would reaffirm this by glancing at his father then his taciturn bride.

Chen guided the visitors in the direction of the long corridor that issued from the functional space. At the far end of the converging sequence of evenly spaced doorways, a square window conveyed a hint of natural daylight. It seemed less a portal than a distant beacon. Barnard studied the open area through which they were being escorted. The walls were bare and dull colored. One offered three tall windows, which were obviously in need of a good cleaning. Most of those in residence sat quietly in the institutional arm chairs with slanted backs, attending neither to the visitors, to each other, nor to the active screen on the wall opposite. Two women, leaning on their walkers dressed in monochrome uniforms of pajamas and robes, seemed to be in sideways conversation. A lone, immobile figure in a wheelchair, age and sex indeterminate, faced the windows. It was very still, staring out with both arms extended out over a low table. Barnard commanded his gaze away, his eyelids tightening to help him refocus his attention on Chen's recitation. She was pointing out the clusters of conical lenses on the corridor's opposing walls, one set over each doorway. Different colors and blink patterns alerted staff, she informed them.

"It make quieta an' fasta respon' than if by room numba."

Barnard took the installation as silly and old-fashioned, but he politely mimicked her look of administrative satisfaction. Rhuum numbah, he mouthed voicelessly, suppressing any sideways glance.

They entered the generously wide corridor, which was home to periodically spaced, classroom-like chairs with fat arms, presumably upon which to support forelimbs for the taking of blood pressure or probing for a productive vein. The odors of stale food, laundered bed clothes, and weakened bodies floated on a base of pine scented disinfectant. Barnard recognized the mix from his hospital visits with Diane. Nevertheless, no specific remembrance of her arose during that tour, only the intimations of inevitability and loss. He peeked at Kyle. There was no hint of a reaction in his straight ahead gaze. He had not been so immersed in his mother's decline as had Barnard.

Ahead, the gray floor presented a smooth, shimmering sheen, with bright patches under each of the recessed ceiling lights. In their walkers, stooped, with their heads perpetually angled down, the residents of Zeven might be disoriented by lines or patterns on the floor, Barnard imagined. A number of the irregularly spaced chairs were occupied. A hand at the end of a skinny arm reached out to touch something of them as they passed, a sleeve or the edge of a jacket. Its owner made an utterance that Barnard did not understand. He could discern only the inflection suggestive of a question or request. A figure came their way, with a distant expression blank as sea ice, using a four-point cane plus an aide for support. Neither spoke as they passed. Close by sat another resident, whose pleasant smile was welcoming. It was a grandmotherly departure from the mask-like stares of the others. Barnard smiled as they approached. Without responding she resumed earnestly addressing the hands clasped in her lap.

A light was flashing at one of the rooms ahead. Barnard glanced at the square frame on the wall aside the jamb – AW-707, with two names underneath – before peering in. He took in the sequence of an unburdened metal cart, a robe draped over the back of a chair, then the feet of a figure lying in one of the pair of adjacent beds and attempted to parse the significance of the alert. After another step, the patient's arms could be seen flexed close, its hands limp. "Man or woman?" he asked inwardly. Next into view came the head, pressed hard against the pillow, with mouth agape. Barnard snapped his face straight ahead. He had seen enough.

The dhup dhup dhup of soft shoes overtook them. An Asian man in gray scrubs entered the room and closed the door behind. The blinking beacon above went dark. Barnard felt a growing unease, a sense of being an intruder. He looked aside to determine if Kyle or his bride appeared similarly affected.

Chen was taking great pains to be thorough. She was demonstrating the level of care in that apical unit as if inured to its impact on the uninitiated. For himself, Barnard failed to grasp the point of the walk-through beyond that of an instantiation of eventualities. Her presentation could have been construed as a warning to Kyle, an invocation of what the future might have in store. Unpleasantries always are easier to manage when remote and still seemingly malleable. If meant to be a homily for Barnard, a validation of his occasional actuarial musings and/or a practical life-lesson, his initial reaction was that it would be preferable to be one of the lucky males who dies before he decays, one who goes too soon rather than too late. Barnard was familiar with physical decline. He did not want to learn about its cognitive counterpart. Cooperation then can neither be demanded nor expected, which increases the burden upon the caregiver. For the afflicted, it is paradoxically lightened. Anticipation only magnifies the pain.

"The trick is not to care," a political hack had offered as advice on how to manage the unmanageable.

Just so; nature does have a way, Barnard reasoned at the time.

"'Tis a consummation devoutly to be wished," he then recited inwardly.

They left Sunset Vista's seventh floor to visit other areas and Barnard never ventured back. Knowing of it was enough. Suppress it, he decided, or let it be escaped altogether. There were ways. The more basic question, one that Barnard's analytical economic mind found difficult to answer later than evening when alone, was how an area as limited as that one floor, Seven, would suffice. How, given the longevity of its guests and the rate of new arrivals, could they not run out of space?

* * * * *

Does he have it?
The recollection of his visit to Zeven makes Barnard wince. He again studies his old friend Frank, tries to judge the likelihood of his graduated ascent to Five or to Six. The familiar questions arise: What would it be like when forgetting goes unnoticed? Would one not know what has become of one's self? Does the I, the first person singular, melt away with neither consent nor protest? These are inherently unanswerable questions. Given the premise, they are irrelevant, suppositive, as are the personal consequents of death.

He releases a long sigh. Reviewing, deconstructing, the recollections of his day contesting sleep, are bad enough. Worse is the reverse, the day confounded by what should remain somnolent visions. The pointlessness of unresolved reflection, of plan absent action, troubles Barnard. He derives comfort from the fact that within this sea of concerns, possibly because of them, there is a welcome island of *ergo sum*. Professor Barnard Cordner is extant. Even if gloomy, the fear of not being sustains the burden of being. Unfortunately, there are few with whom to pursue this. Sallie and Phil, possibly. The tedious Martin Stoole? Not likely. And not Frank. Therefore, Barnard is left to ruminate alone and to excess. It is not what he prefers, but it serves. A chill suddenly cuts into him as he navigates amongst the other attendees. It penetrates with a flash of inert recollection that confounds the present. It must be pressed down, overcome, submerged.

He will defer his chat with Frank, he decides.
He has it?
On the way?
Barnard is here, attending Sundae Nite, mainly for the bounty on the heavy table against the wall. An excess of cloth purposefully hides the legs' cracked veneer, the worn details of the carving. He stands motionless, first to weigh this fact then to push it aside and instead visualize the imminent sweet treat. Scanning aimlessly, he is made curious by the fact of his other friend, Phil Winfree, being in conversation with a partially hidden someone. Neither of the seated pair appears to be paying attention to the surrounding activity. Palms trapped under contralateral armpits, Phil presses his torso against the table's edge.

3 - - PAYBACK

With whom? Stoole? Yes, Martin Stoole.
Shit.

He watches them from a distance. Except for Sallie, it is Phil with whom Barnard talks and spends the most time. Their shared service backgrounds had fostered a friendship that started well before Barnard took up residence at Sunset Vistas. Of the two, Phil's navy life was the more lengthy. He had made a full career of it, saw action in Asia. A few years older than Barnard, clean shaven and pleasant in appearance, he has remained trim and straight. This makes him, also, appear younger than his years. Besides being a prodigious reader, Phil demonstrates an agile memory. Barnard envies him for that. His has a broad range of interests plus the annoying knack of offering plausible detailed opinions on and explanations of nearly any subject, concept, or question, regardless of whether he has or has not any factual special knowledge. He is, in simple words, an occasional bullshitter, as his wry grin often will acknowledge, but pleasant company and never a bore.

Their occasional arguments are often tangled knots, since Phil prefers to maneuver them for their own sake not resolution. Barnard has had to learn to discriminate between Phil's actual and fabricated knowledge, to tolerate the origami of complications and suppositions that Phil can fold from a single sheet of fact. In recent years, after his second retirement and taking on a different role as a local Internet guru, Phil seems to have mellowed, to have become more practical and more grounded in reality.

* * * * *

It did not start easily for Phil Winfree. With no evident career path after high school, he was easily bored, left one dead-end job after another before enlisting in the Navy. Being at sea would be "... a damn sight better than on the ground in Nam," he told his few friends. Aptitude tests revealing capabilities that he had not perceived, his initial tour was a ripe plum, a berth on one of the early, medium size nuclear propelled frigates that served as protectors of larger craft – the aircraft carriers and cruisers. Phil seized upon this newly found good fortune, bore down aboard ship and learned all that the chiefs would teach him about the propulsion systems and their management.

The friendly, accessible European and Mediterranean ports soon gave way to operating in the Pacific, in close support of Vietnam engagements. Phil was focused and quick to learn. He was even tempered, did not wear glasses, kept in shape, and always remained alert to his surroundings. His reward for these admirable traits was to be volunteered for the newly expanded riverine service in Nam. They needed enginemen for the river patrol boats, the PBRs, which were a type of shallow-draft craft totally new to Phil. Water-jet-propelled, fast and maneuverable in tight quarters, these were approximately thirty feet of lightweight target, a contingent of the resurrected "brown water Navy" that provided logistics, med-evacs, and communications. In addition to the nipa palms, mangroves, and dense vegetation of the Southeast Asian swamps and rivers, there were floating mines, Viet Cong, uncountable perils of every description. Besides the snakes, the leeches, the skin and gut afflictions, the reek of jungle and men, the boredom seasoned with wary anticipation, there was occasional close combat. The VC were adept at hiding in the dense jungle along the river banks or in innocent appearing watercraft. While Phil's PBR could return fire, that was not its primary duty. He and his mates were to aid, not engage. He was fortunate not to have been posted to a modified landing craft, an LCM, which were heavily armored but, packed with troops and equipment, were also prime targets.

Throughout his tour, Phil tried to keep hate out of it. Realistically, he could not spend nearly two years amongst over stressed and often over medicated young warriors, twenty some months at the task of boarding and inspecting sampans, which often yielded point-blank small arms and automatic rifle fire, and each day facing, from his perspective, gratuitous malevolence, without acquiring some. Nevertheless, he stayed fixed on the task at hand, kept alert for new opportunities, and focused on the promise of compensatory orders to come, if he survived.

After serving his country in that first of many mismanaged modern conflicts and being one of those fortunate enough to complete their tours with neither external nor internal scars, his earlier test scores plus his experience and quals enabled him to secure one of the new training slots for nuclear propulsion specialists. Until the Navy, it had never occurred to Phil that he was clever. Neither friends nor family had aspired to higher education. At the Naval Training Center, his technical prowess already documented, he proved his ability to grasp the subtleties of nuclear engineering and to master the related mathematics. He was not accustomed to book work, but he dove in, loved every aspect of it – the thick manuals, the detailed blackboard sketches, the challenges. The many months of classroom and hands-on training were more engaging than he had expected. Also, he was not one to dwell on how it could have been different.

3 - - PAYBACK

Never again seeing action, the years passed uneventfully for Phil. He rose in rank, at a steady and predicable pace, to Master Chief Nuclear before retiring. He was barely into his fifties when he reached his thirty and out, used his benefits to enroll in an engineering program. He felt out of place, understandably ill at ease amongst young undergraduates. Pushing that aside, his resolve and credit hour allowances in respect of his Navy schooling enabled him to finish relatively quickly. With a formal engineering diploma plus extensive practical experience, he had no trouble landing a mid-level project management job with an aerospace firm down the coast in El Segundo. While in the Navy, Phil always had been either on patrol, in conflict, or in foreign ports, was always part of some tight team. In Ocean Park, he could stay put and be alone. He could enjoy what the nearby beach communities had to offer.

Winfree still lives in the same bungalow, less than a block from Sunset Vistas, that he bought before the neighborhood's conversion to multi-units and apartment complexes began. He recognizes his insularity. However, an apartment would not work for him. There is his project. Moving would be highly disruptive.

As Barnard was quick to appreciate, Phil is a how, not a why person, an intuitive engineer whose focus is getting *it* done, whatever *it* is. This was an essential aspect of each of his defense related, clandestine imaging and interdiction projects, the last being the drone-based Detect-Identify-Eliminate working group – the clever, unsubtle appellation of a type of "spook work" that precluded outside discussion. This accounted for why Phil came to devote much of his free time to his home workshop and his seemingly quixotic project. It gave him a topic that he *could* discuss.

Sensing the likely eventual decline of his utility as an engineer/technician, in addition to the increasingly advanced capabilities of the younger hirees, Phil signed up for a few economics classes at SMC. He also hoped to learn of some suitable opportunity, regardless of pay, that would counter the expected stasis of the full retirement that was not far off. Phil approached the seemingly engrossed Barnard in the crowded cafeteria one afternoon and asked if it would be agreeable to take the empty chair opposite. Chairman Cordner had needed to get out of his office, needed to get away from administrative details, in order to have an undisturbed moment or two to read. He nodded at whom he took to be a vaguely recognized older student then resumed his perusal of the journal article next to his tray. A similar chance meeting, a week or so later, led to questions from Barnard and a mutual exchange of experiences, which they found to have interesting overlaps: the Navy, where they lived, pressbrot at the OP, walking the beach, and nuclear physics. As it happened, Barnard's focus had been on Diane that day and ripe for interruption.

Barnard came to welcome his unpremeditated talks with Phil. The Economics Department at SMC suffered an excess of monochrome ciphers. He derived little pleasure from hearing their daily gripes, usually petty, the most recent rumor, who was screwing whom literally as well as figuratively. Phil, in contrast, provided interesting conversation. He never felt the need to explore why Phil picked his company for a shared table, if there were any secondary intentions.

When, some time later, Phil described his personal project, Barnard knew, was absolutely certain that it was a put-on constructed of pseudo-science and misstatements of fact. Barnard knew basic physics. It was his undergraduate major before his own abortive Navy adventure. Except for his difficulties with advanced math, Barnard would have gone on for a graduate degree. When Phil started his explanation of cold fusion and the reasons for his interest in it, Barnard arched his spine and pressed against his chair. It was silliness, a dilettante's quest, he grimaced wordlessly. He tried to mask any reveal of his superior insight by feigning interest in the noisy students nearby. Only vaguely familiar with the topic, Barnard yet knew enough to have a thoroughly negative opinion of the alchemists' fantasy that cold fusion represented.

Phil surprised him. He provided quick replies to the obvious exploratory objections. He presented a detailed understanding of the concept, of its implications, and above all, of its small chance of realization, none of which could Barnard fully refute. "Small chance" was, after all, not zero chance. Barnard found Phil's hobby, what he took to be an ill-considered endeavor, nonetheless a stimulating and novel intellectual exercise. They sparred into early afternoon on several occasions. Phil needed an outlet and his optimism was not diminished by the unproductive similar projects elsewhere. His perceived possibility of success, as rationally slight as it might have been, seemed to take the edge off Barnard's insistent critique.

Phil's project was for his spare time. Outside of his work, which he was forbidden to discuss, he had little to command his attention or to convey as his own. Once, when Barnard was newly alone and Phil near to civilian retirement, they met up at the OP Café. Each seeking a break, a confidant, or an encounter, they rode the bus to Venice, wandered among the crowd for an hour or so, then went to Phil's typically Southern

Californian, stuccoed frame cottage, ostensibly to be shown the work-in-progress. Barnard eyed, without comment, the many potted plants crowding the tiny portico. Phil escorted him in under the arch of the congested foyer, into the predictable living room with kitchen and dining beyond, past the short offset hall with the central bath and compact bedrooms at either end. Without pausing over the books and mementoes on numerous built-in shelves, they went out through the dim kitchen, redolent of cumin. As they crossed the short rear yard, Barnard appreciated the shift to a seemingly familiar, fresh smell. There was a trimmed tall hedge along one side of the yard, and numerous colorfully leaved and flowering plants. It was very neat, unexpectedly serene. Barnard looked over his shoulder, back at the compact, low bushes underneath the windows at the rear of the house, then at the grim windowless structure ahead, evidently a converted garage. In the lead, Phil held open the door and motioned for his visitor to follow.

Barnard hesitated. He could make out nothing in the pitch-black space ahead. Then he heard the click behind.

Old-fashioned florescent lights hummed and grew bright. In the center of the space was a substantial worktable, with equipment stacked on it and beneath. Barnard could identify most: DC power supplies, digital counters, a Nixie display stopped at 0739, a voltmeter with its coiled probe wires, old analog and modern digital waveform generators. Looking about, Barnard tried to recall Phil's previous serial commentaries, so that he could make some coherent comment. There were several oscilloscopes, one of which resembled a carted Tektronix of distant memory. Nearby sat a second CRT, with a round, pale green screen and slanted hood, similar to the ancient Dumont that Barnard had commandeered from departmental storage for an undergraduate lab project at UCSB. Near one edge of the worktable was a soldering station. A heavy, black Weller soldering gun sat inboard of it. Elsewhere Barnard saw the typical accouterments of a basic modeling shop – hobbyist's lathe, bench drill press, several grinders. Scratch pads, hand tools, sturdy work gloves, lengths of tubing, and more were scattered about. A black roll-about chest – with an open combination lock loosely through its hasp – stood against the wall. Barnard's bemused curiosity gave way to astonishment.

The core of Phil's project was a stainless steel sphere. It was about the size of a soccer ball and appeared to be of two halves joined along a pair of circumferential flanges. The upper one had a porthole, of thick quartz Barnard later learned. The globe's shiny surface was further adorned with an irregular array of wires, rods, and projecting cylinders, the latter with wires and cabling of their own. As Phil described the intent of the device and its components, Barnard was stunned that the items necessary for Phil's enterprise, a few unfamiliar to him and requiring explanation, were available for civilian use. There were high voltage power supplies, which could kill if improperly used. Barnard was handed coils of precious metal to heft – palladium and platinum. He politely inquired about their cost, about how Phil was able to acquire the heavy water, the several types of lasers, the radiation detectors.

"All available. No problem," Phil told him. "Friends at the base in San Diego helped some."

He went on to patiently explain his setup and the methodology of low temperature fusion. It was obvious he was delighted to have a competent, even if skeptical, visitor.

Discounted and declared foolishly unscientific by the vast majority, cold fusion – the coaxing atoms of hydrogen in heavy water to fuse and thereby release a net gain in energy output at room temperature – was a cottage industry of sorts among out-of-the-way laboratories and amateur scientists. While Barnard endeavored to recall the specifics of what he had read of the controversy, Phil outlined the various degrees of reported success of similar isolated endeavors. The majority opinion – the academic determination – was that the topic was properly relegated to conscientious devotees of the arcane. It was for nutty savants, in other words. Nonetheless, hints of success circulated among the cadre of similarly focused, unattached investigators. Even sketchy anecdotal reports were motivating.

"That's all been shown to be, uh, to be bullshit, Phil. Years ago," Barnard protested. "Sorry. It's one of those crop circle or no-nose extra-terrestrials stories."

"Then why did the Naval Warfare Systems people follow up on it?" Phil countered. "And Energy? The DOE?"

"They didn't 'follow up' on it. They looked at it. Sure, even Schwinger took a look, came up with a hypothesis or two of how it *could* work. Science news crap. That's not proof."

"No. It's not proof. But it's a start."

"Start for what? You're talking about pushing hydrogen atoms close enough to fuse. Right? That's not going to happen except in a hugely hot, high pressure plasma." With that, Barnard jerked his thumb upward.

"Not if they're naked, no. What if they're captured on a catalyst, like really fine palladium powder? Or stuck in a crystal? There are all kinds of biological processes that would be impossible at body temperature except that a catalyst is involved, a molecular structure that torques some protein into shape for the next step. It's a good analogy."

"Analogies aren't proof, Phil. Analogies and could-happens are what keep alien abductions in the news. Right?"

Barnard worked his lips. Impressed by the effort, apart from what he judged to be its misguided aim, he wanted to be critical without being overtly dismissive.

"If you stop calling it cold fusion," Phil insisted, "and use the right name, like Low Energy, or Lattice Assisted Nuclear Reactions, it doesn't seem so weird. Sure, the energy you put in might not push free heavy hydrogens close enough to fuse. Like I said, though, if they're restricted, bound to a catalytic surface or lattice, then the – "

"You can say that, right. But can you give the details? Give a step by step mechanism for it to happen?" Barnard overlaid this upon Phil's words, so that neither heard what the other was saying.

" – atoms can be agitated enough to occasionally close the gap to form helium nuclei."

Phil paused, looked intently at his almost elegant device.

"Yeah, well, I haven't done it; no one has for sure. It's worth a look, though," he then added.

Barnard shook his head.

"Fine. No one's done it. You can, uh, you can throw out hypotheses all day. That doesn't get you very far. You've got to come up with some mechanism that fits with the science we know. Right? Or least show that it happens, actually demonstrate it. Then you have to give people a recipe, a clear methodology, so that they can confirm it. Make it real, else, it's, uh ..."

Barnard could find no suitably friendly words with which to finish.

"Anyway, that's my project. It's kinda fun. Keeps me occupied when I'm not on site. New ideas often pop out of strange places. Often from some loner who's too focused to care that it might be a waste of time, like everyone's saying. Right?"

Phil's sly imitation of Barnard habitual insertion made them both grin. And his intimation of good ideas coming from dedicated amateurs was justifiable. There are ample anecdotes of tyros' persistence shaming professionals' disdain. Asteroids and comets, fossils and obscure life-forms, even mathematical theorems, carry their improbable discoverers' names. Phil subsequently related his favorite example, the discovery of Hanny's

Voorwerp, Hanny's object, in astronomy. A Dutch school teacher, Hanny van Arkel, he noted for Barnard later over coffee, had extrapolated, from a smudge on a photograph, the existence of a new class of galaxy. Ultimately, she was judged to be correct. Hundreds of similar objects were thereafter identified and analyzed for their unique cosmological properties. An amateur marked a path so that experts could follow.

Barnard's concluding comment on that visit was that cold fusion remained an unexplainable phenomenon precisely because it was unachievable, internally contradictory. Phil dismissed that view. The improbability of his project made no difference to him. But he was no fool. Neither was he rash nor eccentric. The fact is, that no matter how far fetched, a project with perceived likelihood of benefit always will be pursued with diligence by a committed few. It is natural, in other words, to hold fast to one's illusions. Cold fusion could be demonstrated one day. One had only to feel the sun's warmth to appreciate that producing net energy from fusion is possible. Being able to do it here, on earth, would make the Muslim Wars, the entire Mideast, finally irrelevant and impoverish the barbaric, oil-fueled terrorism perpetrated in the name of Islam.

Phil knew he was not alone in pursuing his passion for demonstrable cold fusion. He knew also how minuscule was the likelihood of any succeeding. Still, he was inspired by them as well as by those who had found something new in other fields, those who had opened portals of discovery. Barnard, for his part, was as much sympathetic as he was skeptical. Yet, they eventually shifted their talks to other topics, to the financial mess, to VieGie and other Internet enhancements, to how people are misled by each. Therein are illusions, to be sure, but illusions of great practical significance, illusions that both could engage.

*　　*　　*　　*　　*

"Pffhhh."

These snatches of the past have occupied only an instant, far less time than it would have taken Barnard to relate them. There Phil sits, across from the chinless Martin Stoole. For a nonresident, Phil does seem to be in the thick of things at Sunset Vistas. Conceivably, his occasional worrisome asides are prescient and should be attended to. Barnard observes Phil's hands, no longer entrapped, move before him in an accepting gesture, palms up and finger tips almost touching Martin's. They bow their heads. Barnard is embarrassed to be staring. People have many layers, he extracts from the tableau.

Refreshments and conversational reprise are the keystones of Sundae Nites. For Barnard it is definitely the former. In addition to the ice cream, syrup toppings, and chopped nuts, he sees the bowls of chocolate chips and multicolored bits. Still several paces away, he detects that there are new items, what appear to be cereals and broken bits of cookies. He is momentarily tempted by the already plated slices of cake. Wavering, he recovers and scans the enhanced array of sundae toppings.

"Why is this night different from all other nights," he chants to himself. Diane had made that joke long before, when he slipped into bed already naked. He had laughed long and hard when it was explained to him.

His eyes drift casually along the table's length as he meditates on hating crushed Oreos or gummy bear pieces with his ice cream. He has already visualized what he will have. He looks up, past the table, and spots Condi, alone as always. The ice cream, the syrup and the nuts, each of which he can already taste, will have to wait. He walks in her direction.

She smiles as Barnard approaches.

"Professahr Chordnahr," she says with her familiar drawl.

"Hi, Condi. It's Barnard, remember? Just Barnard?" he says to her, his inflection both pleasant and accepting. "Are you doing okay?"

As always, she pays no attention to his admonition. Obviously glad for the visit, she holds up her dessert dish and opens her eyes wide, then tilts her head in its direction. Her face is creased, vaguely leathery, and her dark skin is enhanced by her bright, white eyes.

"Later. I'll get something in a minute," he replies. "How are you today, Condi?" he asks, trying not to focus on her half bowl of ice cream.

He must decide with what topic he should commence, since she has been noticeably less voluble in recent months. While the flow of their brief visits is frequently haphazard and difficult for Barnard to follow, he experiences no lengthy indecision this evening. Condi immediately starts talking about Clifton, her oft recalled one and only, as if leaping past a pause in this or some earlier conversation. Barnard knows the story by heart and carries out his part with a sincere, set smile. The minutes pass. He can float along on the tale, let it proceed as she chooses, since he no longer wears a watch, not since Diane died. It is easier to ignore time that way. The protagonist of Condi's recollection is on the sand at Venice Beach, at the crossbars set up near the doubles volley ball pit, showing off his muscular prowess. Condi is obviously pleased to have the resurrected image before her.

* * * * *

Condi's husband, a Wilson Scholar, possessed a degree from a fine southern school. His curriculum vitae listed teaching awards, credible publications, and several invited presentations. Clifton Carter was youthful, smart, and black – an outlier. After deciding to offer him a faculty post, Barnard brushed aside the expected collegial admonitions. Professor Cordner's choice was not designed to even out departmental demographics. He put Clifton forward based upon accomplishments and promise. Years later Barnard was complimented on his prescience, since Carter made a fine impression as a productive, well-spoken academic and, eventually, was a strong candidate for Vice-Chair. Before that, near the start of only his fifth year, he had come up for tenure. The early review was a reflection of his recognized value to the department. Condi was glad for him and justifiably proud. He hoped, he expected to soon receive the appointment letter that is an academician's eagerly sought for assurance of stability. Denial would have been bitter confirmation of the barriers facing one whose lineage extends deep into the rural South.

It was early December. The Cordners' holiday party was in full swing. Diane wanted Barnard to wear a red Santa's cap, but he had refused. A tiny tree, overburdened by multiple strings of lights, was sufficient. Their uncoordinated flashes of red, green, white, bright blue, and yellow dispelled any ambiguity engendered by warmish weather and the early date. The door to the kitchen was kept closed since Diane, the consummate hostess, favored watchful replenishment of the trays arranged on green felt or red sateen squares in the dining room, den, and bar alcove, over setting out dishes fresh from the oven or stove top. Much preparation had preceded the event, most of it falling upon Diane, as usual. Neither wished to have this exposed by an inspection of their backstage. Cruising around to gauge what needed replenishment, picking up unattended clutter on the way to the dining room, Barnard found Condi sitting alone, in one of the arm chairs pushed against the recently refinished oak chair rail.

"Where's Clifton?" he asked. He enjoyed a sip of his scotch as he looked down at her, explored the smooth dark face with uncommonly European features. She smiled back and tipped her head at the space beyond, toward her husband, who was talking with a pair of students.

Barnard's eyes drifted lower. Reluctantly he averted them from the brown curves and shadowed crevice to Clifton beyond. Well done, he noted silently, only in part meaning Carter's easy manner as he gestured and talked.

"I'm glad you both came," Barnard offered to her. "Having a good time? Have you met everyone?"

"Oh, yhes. Whe ahr, yhes," she said slowly. "Iah've bheen on mah fheet awl dhay, so ..."

She lifted her empty plate, wine glass neatly centered, and set it down again upon the ripples of her dress, which was pulled tight across her thighs.

"I understand. I could use a sit-down myself," he said, without changing his stance.

"Sit heah, thehn, foah a whiale." Her warmth was sincere. She paused after her invitation. "Iah dhu whant ta thank yhou foah inviting us. It's bheen verah nhice tuh meeht everah whon, of khourse. Bhut also Iah whant ta thank yhou foah putting Clift'hon up foah promhotion. He has woirked real haard."

Barnard still standing over her, she patted the seat of an adjacent chair.

"Right, Condi. He's done very well. I've, uh ... I've got to, uh, make sure, uh ... For just a sec, then," he finally acceded.

Barnard was glad for the respite, for her steady gaze of invitation. He felt a surge of pleasurable relief in his legs, and something more. It would be an opportunity to learn about Constance Carter, he rationalized; he knew much less of her than he did of Clifton. As they chatted, she revealed where she came from, how she and Clifton met, what she did before marriage, that she enjoys her job. Barnard has retained little of that conversation, except that she worked for the city, and that she and Clifton made good use of the opportunity to travel in the summers. Barnard can vaguely recall inviting them to go sailing, which never happened. Nevertheless, on later occasions, at odd moments, he would find himself imagining how the stately Condi might have looked in shorts or a trim swimsuit.

That mild coastal winter night, she leaned toward Barnard as they talked. The chairs were close and their knees touched. He imagined he could feel hers pressing against his. Her scent was notably different from Diane's, a polar opposite it seemed. Tendrils of it drew him in. His habitual formality was greatly diminished by the two, possibly three single malts he had already enjoyed. He felt an easy suffusion of well-being. Suddenly, he sensed it was time to disengage.

"I need to, uh, to check on things. Busy host. Right?" Barnard said down to Condi as he arose, being careful to regulate his gaze. "Don't sit here by yourself," he suggested. "Have you met Lauren Ketchum? She's Peter's wife. And Everett?" He motioned her up and made the necessary introductions. "Pardon me," he said, once he sensed he could exit the mutually engaged group, and rejoined the general bustle.

Judging from the rising level of chatter in their living room, Barnard imagined that the liquor might be running low or that beer might need to be pressed into the ice-filled picnic chest. These were the excuses, at any rate, for remaining in motion. There was an unopened bottle of Everclear and plenty of juices, so the punch on the dining room table was not going to run out. A quick survey confirmed that the wine and mixers seemed ample. While the tequila had hardly been touched, the other bottles were receiving such fine attention that he might need to fetch more. He replenished his Lagavulin before picking up a plate and assessing what remained of the food. It was approaching eleven o'clock; not late, even by mature academic standards. He felt he had visited around sufficiently. Nonetheless, he took note of those still present, and of his recollection of the few who had opted for an early departure, to make sure. The clusters of guests seemed lively enough. Topical issues, local or distant, major or trivial, were apparently sufficient to keep the guests actively engaged. Luckily there were no raised voices, no political-cum-philosophical debates that required gentle modulation. All in all, Barnard was able to judge that it was a fine party.

He made another sweep of the room for empty snack plates or errant napkins, for tumblers or wine glasses without a coaster or at risk on the arm of a chair. Otherwise unobserved, Diane smirked at him over the shoulder of Dan MacDonald, the venerable former chairman, who was part of the group encircling her. Barnard smiled back. He appreciated that she was the more adept at these evenings, more amenable to casual conversation. Afterwards, while they cleared and rinsed, she would fill him in with the gossip and observations than he had neglected to capture.

He looked beyond the broad arch of the dining room entry at Clifton and Condi, plates in hand, as they studied the ample savories and snacks. He took another sip of scotch, watching, over the rim of his glass, the movement of her upper thighs as she gestured and selected, appreciating her athletic calves and trim ankles.

"Pass these, Honey, would you? They're hot, too."

Diane held the tray out for him. Barnard stiffened, mildly embarrassed, and gave her an affectionate kiss on the cheek before relieving her of it.

"It's going fine, right?" he grinned at her.

"Yes, I believe it is. One of us should visit with the Carters, don't you think?" she suggested.

"I did. Earlier," Barnard assured her. "They're doing fine as well, Sweet. I've been watching."

"Not hard to do," teased Diane.

Barnard made no reply. He kissed her other cheek before moving forward toward their guests. Condi shook her head to convey an appreciative but demure No to Barnard's raised eyebrows and the uplifted tray.

Mentioning the table's array and the plentiful drinks to each he passed, Barnard made another slow circuit. One way to tell if an evening was winding down was how much attention their guests were paying to the food. The nearly depleted bowls and platters suggested that it would be thinning out soon. He was gratified to confirm having hosted another pleasant event. He paused by the punch bowl, to estimate what remained of the covertly powerful concoction, then entered the kitchen and set the tray on an empty section of countertop under a window. Staring out, past the reflection of himself and into the darkness beyond, he heard Condi laugh behind him.

"Whar yhou whorr'd therah whon't be enuf puhnch?" she asked in her breathy voice. She was as fetching in manner as she was in appearance.

Barnard turned about. "No," he said with a relaxed smile, "not worried at all. There's plenty of fixings here. Besides, everybody's nearly at their limit. Right? The party okay, Condi?"

"Whonderful, Professahr Chordnahr."

"Barnard, Condi. No need to be so formal," he advised pleasantly. "We're not at the office. You and Cliff are our friends. Call me Barnard. Right? Or I'll start calling you Constance!" he added with a laugh.

"Whell, Iah jhust whanted yhou ta khnow. Cliff, Clift'hon and Iah, rhealah appreciate what yhou've dhun heah, at thuh school, foah him."

She bore down on her enunciation of "appreciate." Otherwise, she tended to give free rein to her strong accent. It was an aspect of her charm, one enhanced that evening by the potent punch she had enjoyed, he surmised. And by his own, often replenished tumbler of single malt.

Barnard shook his head to accept and at the same time deflect her gratitude. He picked up a large bottle of blended fruit juices with one hand and that potent, nearly 200 proof secret sauce, which he intended to add to the remaining punch, with the other.

"Look," he said. "He's good, very good. He's energetic, productive, and engaging. He'd do fine anywhere. We're, uh, we're glad he's here."

Her eyes, nearly even with his, were intent and steady. She put her hand lightly on his upper arm. His biceps was tense from holding the Everclear.

"Iah dhu so whant him ta dhu whell, ta geht established hea. Iah whould dhu anything Iah khould ta make that happen."

Barnard could not meet her firm gaze for long. It was easier, amazingly, to guide his eyes on a slow cruise down her neck, along the youthful trimness of her shoulders and breasts.

"Anything. Iah. Khould."

Barnard blinked a few times, slowly, feeling her fingers through his jacket. His groin tingled with a familiar tension and the front of his pants seemed to constrict. The two bottles were welcome burdens. Barnard shifted his weight to break his gaze.

"Let's go back, Condi," he suggested. Taking a step sideways, he tried to acknowledge nothing, to reveal nothing. "It sounds as if they're having too good a time out there."

He pushed open the swinging door with his backside and watched her join a cluster by the window.

Barnard never doubted that he had made the right choice, even though he occasionally fantasized the alternative. The tempting encounter was so fleeting, its outcome thereby so malleable, that it allowed for a wealth of virtual realities.

* * * * *

Barnard refocuses on Condi's southern inflections as she approaches the climax of the familiar Clifton tale. Her smile, her splendid eyes shifting this way and that, signify the genuine significance of the replayed past. Before she reaches the point where he of fond memory finishes the last of the grunting pull-ups, a personal record, it has always tickled Condi when Barnard would ask, "And did he hurt the next day?"

She has grown heavy, aged quickly. Barnard is glad that the southern coast nevertheless has remained evident in her speech. Her accent is unique at Sunset Vistas and obviously not nearly as pleasant to others as to him. His eyes shift to the nearly empty dish, precarious on her lap, as he puts the question.

"Hoirt thuh next dhay? Ha. He khould hardly steehr thuh khar!" she starts.

Her focus seems to waver. Barnard continues to smile, linking his recollections to hers, watching her expression flatten and her eyes lose their lock upon his. Clifton had barely reached fifty and long before had been forgiven his trespasses, as she had hers. The culmination of that episode, on the bright SoCal day when Clifton had done his best to show up the muscular hunks who were so serious in their own efforts, was later getting a great deal of mileage from his expansive elaborations on the event, showing no sign of the clogged left circumflex that would kill him the following Sunday after an oversized breakfast of upstanding eggs, thick ham, and crumbly biscuits covered in peppery white gravy.

Reliving the details as well.

Barnard exhales into the space to one side of Condi.

"Can I get you another sundae? A small one?" he then asks, anticipating the sweet, cold crunch of his own construction.

They both know she cannot. In agreeable dissent, she rotates her head left then right. One proper dish of Graese's Old Fashion Farm Vanilla Bean Ice Cream is enough sugar for her.

Nuts? God, no. Absolutely not. Never nuts.

Condi stares straight ahead as Barnard stands. He looks down at her. The barely aimed query was the proper and considerate exit line. He will see who else is attending, then enjoy his sundae with Sallie and May. After pausing, as age often mandates, to be sure that his stance is stable, he steps away, past the young visitor who is likely to have spent a bored six minutes with some relative and now, in the midst of placing his emptied dish in the tray waiting on a rickety stand, is no doubt anxious to leave. He is familiar somehow, but Barnard cannot place him. He tilts his head and receives a noncommittal head bob in reply.

Hurry-scurry off. Right. What up to next?

Something cues Barnard to locate and thereby anticipate Doug. He does not see him. He notices, instead, that Phil and Martin have been joined by Julie Chen. The mismatched pair and she are in frequent conjunction in recent months, either in her office or conferring in one of the common areas. There are neither plates nor dessert bowls before them this night. Barnard is amused but not surprised that Martin and Phil are not indulging.

Neither is a big eater of sugary desserts – Phil because of his incipient diabetes, Martin because of vanity tinged attention to his waistline. Chen's abstinence this evening is less easily explained. She usually has a plate or coffee cup in hand when in attendance to chat up the residents. Her forearm rests upon several closed binders. Something must be going on, of which he concludes he is not aware.

Martin Stoole is tall and lean, at least a quarter century younger than Barnard. His attempt to deflect attention from his absence of chin, by sporting a large, dark, hair-comb mustache, is insufficient. The intimation of weak character remains. Barnard had hired him, thinking that the Economics Department would benefit from an instructor more attuned to the students. It was soon apparent that Martin Stoole, a plodder neither bright nor dull, a seeker of superficial seams of knowledge to mine, a picker of low fruit for regurgitation, would never be more than a journeyman academic. Unfortunately, policy and budgetary restrictions conspired to allow Barnard no latitude for seeking a replacement. His administrative remorse at the commitment was partly to blame for him taking advantage of Stoole during the new hire's second year. An example of the law of unintended consequences, Barnard's subsequent, guilt imbued overcompensation worked in Martin's favor, facilitated his obtaining tenure.

Martin is most notable at Sunset Vistas by virtue of his Current Affairs series. Generally less discussion than lecture, each of these hour-long evening meetings is an accessible alternative to the manufactured thematic uniformities of the daily news shows. Likewise, the Internet offerings are often so slanted and disingenuous that only those hopelessly shallow or impaired fail to recognize that additional points of view are needed. Martin is appreciated by the retirees for attempting this, for seeming to address their specific questions and concerns. He and his invited speakers keep their aging minds attentive, provide them with opportunities to dwell less on the past and to engage the present. While Barnard occasionally attends, he normally does not take part. He prefers to be his own arbiter, to either extract hidden kernels of truth from the various streams of information or to disregard the whole of them. He favors his own haphazard searches, his presumably unmonitored discussions with Frank and Phil at the OP Café.

Reflecting on Martin Stoole commonly leads Barnard to consideration of the actively touted VieGie. From this it is but a short step to consideration of Elder Edens. Barnard is prone to take issue with

VieGie's utility as well as with the Elder Edens' practicality, while Stoole promotes both at every opportunity. Barnard does not wish to be drawn into any such debate this evening and senses the wisdom of avoiding the mutually engaged trio. Therefore, he redirects his attention to the augmented array of Sundae Nite treats. Now standing at one end of the long treat table, he studies the stack of faintly embossed ice cream bowls, with napkins adjacent. He sees the forks for the cake but does not see the spoons for his ice cream.

Where spoons?

Large glass bowls are pressed upon a mound of ice, which Barnard notices is melting rapidly. The ice cream must be softening, he fears. Yet, while it has lost shape, it does not appear inconveniently drippy, as often happens by this time.

Why so crushed today? Melt too fast. Ahhhh, right. Takes heat away fast. Add salt, keeps even colder.

Right.

Servers understand that? How? Where from?

Barnard is visited by images of Father, of their prematurely cut off outings. He recalls the ice cream shops on Montana and on Third, of pleasant evenings out alone when he was ten, with a little extra money in his pocket. Two scoops were so fine then. They still are. He is going to have both tonight.

Right. Chocolate AND vanilla. Right, right.

"Spoons?"

His interrogatory is softly voiced. He looks down, then toward the far end of the table. May and Sallie, nearby, have served themselves. Their exploratory looks signal that they, also, are dealing with the apparent omission of spoons for their desserts. Sallie and her ever present May are mobile, agile even, happy to fend for themselves. Like Barnard, they are not going to ask. Unlike Barnard, they are not going to fuss over it. They are momentarily screened from Barnard by other attendees as he starts in their direction.

Say hello. Let her see I'm –

Oop. Shit!

Sallie and May are again in Barnard's view, spooning up portions of ice cream. Unfortunately for Barnard, Doug is close by, is perhaps attempting to chat them up. His nephew is at the table, exploring its array as well and apparently for the second time, Barnard thinks back and concludes.

Doug. DAMN it. And it WAS him before. The nephew. Look-alikes. Peas in a pod.

Barnard would prefer to make his sundae, then to chat with Sallie and May as he enjoys it. He chooses, however, to wait until Doug has left or, at least, until he is not near the two ladies. He steps back and looks about for whom else he can visit with. Nearby is Emmanuel Konrad, whom he recalls from high school and who is definitely on the way to Five. Manny accepts help, seems to be ready for that move "uptown." Manny, who was always poking. Manny, who always found a seam, a way to needle. Manny, the wannabe-jock with sharp elbows when they played basketball in the stuffy gym with Jer, Larry, Big Terry, and Little Terry. Barnard never looked upon him as a friend.

Judith. Wanting to, close to, until Manny. Lying bastard. Screwed me up with her. Awful. Never would go out again. So close.

Probably needs help to wipe. Good. Next stop ZZZ. Ha! Piss on him.

So close.

Barnard dams up the stream of unpleasant recollections, forcing his glance to the demure, never-married McConnally sisters. Twins, identical from hair to mid-heeled shoes, they are perched on chairs angled in toward each other. They were already in residence when Barnard arrived but he has not come to know them. "Surely a kinder, higher power will provide that they die together," many have said. While Barnard has come to appreciate the sentiment, he finds it devoid of practical relevance.

He stands aside as Barbarelli squishes by on his wheeled chair. The glumly intent, elderly Italian looks up at Barnard who, to be perverse, presents him with a broad grin before resuming his examination of the assembly. It is a collection of largely disparate people, most of whom are residents and familiar but little more. Virtually all – whether quiet or animated, alone or with visitors – are, as is he, in their putative golden years.

Goldene Jahre sind Glückliche Jahre. Right, golden.

"But for whom?" he asks aloud after another lengthy exhale.

There are fewer people here tonight, as Sallie suggested, fewer visitors, in particular. Barnard imagines that the change to hotter afternoons is partly the cause of the latter. Since VieGie is so entrenched, those who would come, for a few hours or a fraction, to keep in contact with their familial elders via its remote visitation capabilities, perhaps find it an appropriate concession to their busy schedules. The analogous question is: Do Grandpa, Aunt, Father-in-law, or Mother share that sense of convenience or merely accept it? Probably a soupçon of both, he has already decided. Also, some of the residents, who would normally come under his gaze today, may have already relocated or, in light of their reduced capability, moved to one of the upper floors. The same could be looming over his own future without his being cognizant of it. Regardless, Sundae Nite, Barnard reaffirms, is neither the time nor place for these dispiriting considerations.

Sundae Nite draws him in. Barnard takes advantage of and appreciates the small pleasures it provides. As always, there is an additional subtle benefit. The event tempers that fear of the inevitable end that all seek to delay. Witnessing the drama of gradual decay in others makes recognition of its alternative – a quick demise, which is the way he would have it – palliative and, paradoxically, even protective. Barnard would not presume to utter Hamlet's existential concerns, although he feels them. He steps back to the table, grasps the thin-walled plastic bowl firmly, to steady it, and ladles in dollops of as much creamy white and rich brown as it will safely hold.

Soft. Syrup over.
Right.
And nuts. Colon pockets? So no nuts? No. Must nuts. Must.
Such little bowls.

He crisscrosses the mass with several passes of dark syrup and adds one, then a second helping of the chopped peanuts. Belatedly, he retrieves his customary pair of napkins. One will be for his lap, the other to wipe lips and chin. He resumes his search for a spoon, the spoons. Growing impatient, he knows they must be nearby.

"Pfffffffhhhhh."
They ... got ... them ... Where? Ahh.

He follows the lead of his eyes toward the pink plastic utensils he at last has located, nestled on a cloth at the back of the far end of the table. He shakes his head slowly.

Why cloth? They're disposables.

Barnard sets down his construction to await his return from the necessitated short trek. Arriving at his goal, the two napkins clutched firmly in one hand, he leans across to pick out a spoon.

"Should be together," he mutters, his cheeks warm as if from shooting a high tropical sun. "Silly. Plates and bowls there, forks there. Spoons here? Stupid."

His sotto voce complaint, he fears, is certain to elicit another nocturnal monologue or the composition of another never-sent Memo to Staff. He again approaches the figure who, now lounging near the middle of the table, is forking a slice of cake. Disjointed images blossom and overlap as in a kaleidoscope as he steps around. The present dances with the past: snatches of Sunset Vistas of this day; of how it was, of how he got there and why; pieces of ancient events, of what may yet come to be. It takes barely an instant for him to skitter through decades.

I'm here. It's now, damn it. Now.
Not declining. Won't. Like a young sixty-five.

Barnard walks back slowly, deliberately. He blinks to clear away the chaos of competing allusions.

More reps. Tomorrow. Need arms strong, quick. Explosive.
No reason, except ...
Wish slept better. Too much sitting.

He takes a closer look at the casual form in passing. Experiencing a stab of annoyance because of this second, obligatory detour, he sucks his tongue as he transfers the adjudged to have been administratively malplaced small spoon to join the napkins in his left hand. Barnard pauses, feels a slow, deep breath expand his chest. Looking aside at the teasingly familiar figure, it seems he is having ice cream, not cake. Barnard stands immobile and drifts into an intentionally disparate reverie of anticipation.

Doug! Again!

His awareness of place is overtaken by the unwelcome visage, one appearing slimmer, fitter than it should. It is the ever annoying Doug who is lounging by the ice cream, his ankles loosely crossed, his haunch pressed against the table edge. Barnard considers whether to acknowledge him as he walks past, whether to offer a verbal greeting, but quickly dismisses both options.

Spoon and napkins clutched in his off hand, his right in the midst of extension, he stops. There is no treat. There is no sundae with chocolate, vanilla, syrup, AND nuts awaiting his return.

"What the hell!"

He hears his declamation repeated, louder the second time. Could they have cleared so quickly? he asks himself. There is that verity of instant replay, the seemingly incontrovertible fact of having left the treat behind to get a spoon. The various possibilities unfurl in a cascade of images:

It was knocked off.

No. No spill.

It was cleared by staff.

No. No one walking away.

He never made his sundae.

Oh, Christ, don't be that.

I remember, I remember.

The listing figure beside him slaps a dessert bowl down hard upon the table. It is loud, that thimpt! of sharp contact of stiff plastic upon thin cloth over European hard wood. The impact disturbs the clinging white and brown residue, the nut bits imperfectly glued to the side of that bowl by syrup.

His bowl.

Barnard's mouth opens slightly, a tentative, silent fish gasp. Then color floods his face. He feels its searing heat.

"That was my sundae," he complains firmly. "You saw that. Right? You saw me put my bowl there. Why did you take it?"

The figure looks to the side, then directly back at Barnard, presenting a tight, sardonic grimace.

"What're ya starin' at?" it grunts.

Barnard has no idea of the why of it. He struggles to understand. In a flash he simultaneously grasps the figure and the meaning of the smirk. Barnard knows, as once he knew.

Douglas!!!

"That was a rotten thing to do, for chrissake," Barnard says, still in an unabashedly loud voice. He is emboldened by his familiarity of place, by this Sundae Nite, by the many Sunset Vistas Sundae Nites of the past and of those still to come. At the same time, he feels dislocated in time. He waits for some reply, anticipating "I thought it was mine," or "I didn't see you leave it," or "I figured you didn't want it," or, even as unlikely as he rationally knows it to be, "Sorry."

"Ya're still here, old man. Why?"

"Still here? Certainly I'm still here," Barnard feels he must reply, his eyes open wide. "Why shouldn't I be?"

"Because ya're old. An' ya're Barn Yard. Who wants an old Barn Yard hangin' 'round?" the figure states, the prolongated affront precisely phrased, blithely rhetorical, barely lipped.

The figure's eyes are at a level barely above Barnard's. It is Douglas staring fixedly back at him, as if summoned up from a distant past.

"No. I'm not old," Barnard spits out.

The figure's head tilts forward and he sniffs twice, as if testing the air, sampling the reality of the Barnard before him.

"Yep. It's Barn Yard."

As rehearsed in fantasy, many times since taught as one step beyond sparring, Barnard moves. Felt as potential necessity in midnight remembrances of conflicts escaped or denied, this time it is real. With a quick extension of his arm, drawn from imaginary battles and choreographed rehearsals, the heel of Barnard's free hand crashes against the nose above the smirk. This time he hears it.

Craack!

Barnard moves in a half-step. He loops his other foot behind the lazily crossed ankles and jerks them toward him, breaking the perch so that the Douglas slides downward, the back of his head hitting the table's edge with a hollow thunk, both hands en route to his face. The effective maneuver is the low leg sweep that Barnard had taken pains to acquire long ago, after being embarrassed by a much younger opponent's takedown. The figure is making that "uh uh" noise – the expected short, rapid gasps. Blood will appear, of that also there is no doubt, but Barnard has already turned away. He will not see it. Nor will anyone look at him, he intuits. Everyone aware will look to where he was, to the table with the ice cream treats, to the cake and the suspect coffee, to the splayed legs on the floor. There will be no rush forward. Residents do not rush, they gather. Rushing is for staff. Riding easily on the current of an ebbing tide, Barnard neither looks nor sees nor hurries. The room is hushed. His ears are filled with the roar of a cold, engulfing sea. He feels its embrace.

Well, no more Sundae Nites.
"A Credit Nation," is what Julie Chen had said.
Right.
That's how she ...

CHAPTER 4

WHO IS BARNARD?

Barnard's schoolboy reticence, inherited neither from A.G. nor Father but probably attributable to the former's overbearing manner and the latter's premature passing, made books fine companions. In high school he had shown mathematical aptitude and a keen appreciation of science. He was enthralled by his teachers' presentations of easily appreciated mysteries and how these could be explained and quantified: the paths of colliding billiard balls; multiplication of a man's strength by levers and pulleys; heat moving through air and sound through water; what motors and generators have in common; how curved, polished glass altered our status in the universe; why ice floats on the liquid sea and does not sink to make of ours another hostile, frozen planet. Revealed to Barnard was a world that could be more than experienced, it could be understood and, therefore, controlled. Well before his senior year Barnard chose physics as a career.

At the Santa Barbara campus of the University of California, not far from family and friends, he was immersed in more profound concepts: electromagnetic waves – how they arise, propagate and convey information; the innards of molecules, and the innards of those innards in a stepwise descent into the incredibly infinitesimal heart of space itself; stars, spiraling galaxies of stars, groups of galaxies coalescing into clusters hinting of a humiliating structured immensity; quantum mechanics, with its inherent indeterminacy and the illustrative conundrum of Schröedinger's cat. It was at this juncture that doubts arose. Learning to manipulate concepts relevant to a world that he could experience was thrilling. Being humbled by the abstractions of those extremes that he could not, was not. Barnard began to struggle.

Henry and David, his closest friends, understood with ease what was increasingly frustrating for Barnard. There were hours of study and explanation. How, for example, could a particle, arbitrarily distant from a co-created partner, instantly mirror the other's change of state? That "what is" is influenced by his observation posed no great problem; he could manage both Heisenberg's uncertainty and the double slit experiment. In contrast, that "what was" might depend upon "what will be" was totally disorienting. Evidence for the paradox of retrocausality was yet to come. Even if this had been there to consider as experimental fact, time's flow was too concrete for him. Surely, cause cannot follow effect; that would beggar the meaning of the words. The sterile juggling of abstract symbols provided him no insight into the underlying realities. It seemed to impede it. He felt the frustration of a strenuous hike in the wrong direction.

"What do you mean, what's it mean?" Henry said after circling the squiggly symbol still in place near the center of the blackboard.

David was off to one side, his buttocks pressed against the back of a chair. It was a late fall afternoon. The three friends had watched the last of the other juniors follow the professor out and were alone. Barnard had leaned his head back and closed his eyes after the hour-long lecture on the utility of eigenvector analytics before posing his question.

"It's an it. It doesn't have to *mean* anything. It just is," Henry stated.

He pushed a stub of chalk diagonally down through the neatly aligned rows and columns of the subscripted X's of the matrix on the left.

"It's there to stand for converting this," he said, tapping the array and his newly added white line with his knuckle then waving a circle around the whole of it and concluding with, "into this."

He pointed to the other matrix, the pair of large brackets on the right, which enclosed only a staircase of subscripted capital A's running from upper left to lower right in, to Barnard, vague sympathy with the overlaid slash on the left.

"Everything else in here," Henry said, "everything off the diagonal, is zero. All zero matrices, to fill in the blanks. So that these guys," tap tap tap on the diagonal of A's, "are square, like he said in lecture." He paused, looking intently at Barnard and anticipating, hoping for some sign of enlightenment. "It's a symbol," he insisted, with mild exasperation. "It's a symbol for the conversion."

Henry swept his arm from left to right several times then underlined the intervening mathematical glyph with repeated heavy strokes of the chalk piece, emphasizing his point with a sharp smack on the black slate. This was another of the many times he had tried to help Barnard. He hooked a thumb behind his belt and struck a pedagogue's pose.

"It stands for the transform of this," he iterated, smack left, "into this," smack right.

David looked on quietly as Barnard stared at the analytical shorthand. He had barely entered the labyrinth. He knew the mathematics ahead would be increasingly complex. Its physical manifestations would forever elude him.

"I understand what you're saying, Hank. I just can't picture it," he sighed. "I can't see how the fuck it works."

"No fucking need, Cordy. You don't need to *see* what it means or how it works. It just damn *is*, damn it," he laughed. "It's a trick, a useful t-t-t-tool."

David shifted his stance and tightened his arms across his chest. "Like Hank says, you don't have to dig into the theorems that prove it works. It works. The math's a tool," he echoed. "Whether you can picture it or not doesn't matter one flying fuck. Just use it."

He was, Barnard thereafter concluded, unlikely to be a successful theoretician, an insightful innovator. His grasp of the inner mysteries was doomed to be forever inadequate. Physics as a profession could yet be rewarding. He was clever, perhaps even smart. He could take the path of applied or experimental. He could design, build, then run the complex assemblies of equipment that others had conceived. Intriguing aspects of the world's underlying mechanisms would be there for him to explore, if he would submit to being dependent upon the minds of others, to being in support of their endeavors not his own. This picture of forever being ancillary discouraged him. Only later did he come to appreciate that to follow is an honorable role. Only later he did come to appreciate that in most pursuits meaning is subservient to utility and that most who pursue have actually been led.

Our experience is conditioned by the world of physical action. We are bound to what our senses present. Few will retrain their brains so as to achieve understanding that leaps past the associated constraints. Only the truly gifted will go beyond the sensible into the realm of the unimaginable and make a coherent picture of them both, will extract from the mathematics a communicable reality. Barnard had reached an intellectual boundary that he could not cross, a dense bramble that he did not dare enter. It was beyond being smart or dumb. Failed priests suffer that moment. They may desire, they may even sense in an extremely basic, sincere manner a truth that is beyond them, yet not have enough strength to submit to it. It was an apt parallel to his dilemma.

The difference between religion and physics, the undergraduate Barnard was obliged to admit during one of their off-campus sessions of bibulous argument cast as tutelage, consisted of the testable predictability inherent to the latter and dismissed by the former. Barnard tap danced his again drained beer mug over the table. If, he offered vainly, the physicists' predictions derived from concepts that could not be grasped, then it must reduce to the same dependence upon authority and accepted interpretation as with scripture.

"Right, Hank?"

Barnard was not sympathetic to a priesthood of theoreticians, whose visions he would need simply to accept. He had little interest in a career that required acting as their acolyte, their handmaiden.

"The trick in quantum mechanics, Cordy," Henry oozed out, "is not to go looking for reasons. If you want to really know how some of that shit happens, you'll go batty. No one's smart enough to do that. It just fucking does. It's just what shows up when you do the fucking experiments. If the math works, then wipe your dick and zip up! That's as good as it's gonna get."

"Hank's right. Reasons are like smoke, Barn," David put in. "They're there, but there's no point trying to grab onto them."

"What if they start running the world that way?" Barnard posed in a gloomy tone.

"They won't," David said with semi-serious finality. "Not smart enough. Never will be."

Barnard shook his head.

"If quantum mechanics is fucking right, than everything that happens is fucking predetermined anyway," he ventured. "Right, guys? So no point."

"Yeah, right," laughed David. "No point. You're exactly, fucking riiiight!"

Henry, who had been sliding his drained schooner back and forth on the wet of their table, suddenly sent it dangerously close to the table's edge, daring it to fall.

"I'll never get it," Barnard lamented.

"So what, my hyperdimensional, twit bud," offered Henry, whose future was well defined. He looked up and out over the noisy Friday night throng of sleep deprived students and employables starting their two days of reprieve. "We're still the smartest fucking t-t-t-table in here." Laughing loudly, he beat a brief tattoo on the table top then waved the waitress over.

Before the end of that year, Barnard acknowledged his loss of motivation to go on to graduate school. His advisor's patient championing of the valued role of the implementer, of the facilitator, was to no avail. Mother would have been delighted to press her sister for his insertion into the business of the dairy. That would have meant changing his major to Business and, after a suitable interval of apprenticeship, dutifully acquiring his MBA. This was far from an exciting prospect. Nor, he knew, would he be able to tolerate his Aunt Laura's condescension.

Moving out of the frat house was a dramatic change. He was taken aback by how strong his feelings of attachment had grown. Staying, however, would stifle the necessary reconsiderations. He found a compact set of rooms not far from the shore, on an upper floor and oriented so as to capture some of its sounds. The occasional screeching gull, the faint, dull whumps of the surf, the rush of the often heavy breeze past nearby palms blended with the steady buzz of motorcycles and the hum of trucks on Shoreline Drive. A previous tenant had left behind a redwood lounger on the patio downstairs, of which Barnard made good use. When others, inland, were shading their eyes from the sun, he would recline in the calming milky white gloom of midmorning mist to study and to intermittently sketch out alternative futures.

During the summer break that followed, preferring to stay in Santa Barbara, Barnard spent much time on single-mast beginners' sailboats, usually a Sunfish. The university's sailing club was informal. He could select from among their small fleet as he desired and gain skill at his own pace. He sailed the Channel, navigating the often active currents among its islands. Being alone helped him to focus, to plan. He ventured beyond the sound of pounding surf, learned to pay attention to wind and weather, to make the minor observations that can presage major change. Night did not bother him. Neither did fog.

Once, with familiar landmarks impossible to see through moist low cover, he so lost track of his course that he got carried miles north. Like a moth, he eventually was attracted to a fuzzy, blinking light that marked the entrance to a basin hosting an untidy collection of light-duty sailboats and small power craft. It was several hours before the surface-hugging clouds blew off and he could sail back to Santa Barbara Harbor. Being disoriented, being delayed, being alone – he had not been discomfited by any of it. He had not taken into account that dangerous rocks could be a few meters away without his knowing, that a sneaker wave could be upon him at its whim. When the fog lifted, these dangers could be seen. They would necessitate only a minor adjustment to avoid. Safely berthed and back in his apartment, he began to think about that. Gradually he came to appreciate the lesson that had been put before him: A plan based solely upon knowns is half a plan.

In a heavy sweater against the chill, Barnard lifted his head from the arcane semiotics of Chapter VI ("Application of the Calculus of Variations to Eigenvalue Problems") in Courant and Hilbert. It was the winter of his final year. He was still struggling with parts of that two volume set he had borrowed from Henry, the by then recognized genius who had absorbed its entirety, the one for whom the dense material was "...ch-ch-ch-child's play." He stared out across Shoreline and attempted to put into concrete form the decades ahead. Without a graduate degree, the best he could anticipate was a post as a junior engineer, a lab associate, or far worse, as one of a host of apprentice project managers pushing paper at some defense contract company. The last of those potentialities presented a positive side. It might keep him from being called up. Suddenly, he had an outlandish idea: Why not the military as a career? The navy, specifically. He enjoyed sailing; he enjoyed the sea. The Vietnam War made enlistment preferable to succumbing to the likelihood of conscription into ground combat.

Frances was not happy with her elder child's incipient decision. She took any sort of military service as an assignation with death or worse. In contrast, the Navy Chief, a submariner with whom he spoke on an early Career Day, was eager to respond. He outlined the new Nuclear Navy, which would offer the soon to graduate physics student an exciting field of specialization. Barnard's attention began to wander. Playfully misconstruing the petty officer's title, he auditioned the husky man with the thick neck for a new role, standing before a tall stockpot and wearing a white chef's coat. The false image faded rapidly as Barnard felt himself drawn in by the relevance of the chief's commentary.

During the follow-up meeting that same month, in an office at the rear of the Naval Recruitment Center, Barnard heard a potential future unfold in detail. The commander diligently and carefully informed Barnard, patiently answering his questions while simultaneously evaluating the prospective enlistee. There would be three arduous months of Naval Officers Candidate School, he said. This would be his basic training: discipline, duty, obedience, the service. Then would come a year or so of focused schooling.

"It'll be tough, but you'll handle it," the officer predicted, speaking as if Barnard's course were truly laid out.

The exact nature of his subsequent three-year posting will depend upon which nuclear service he chose – surface or subsurface; he will not have to make an immediate choice. After a few years at sea, he will be moving at flank speed up in rank, Barnard heard. He tried to gauge the character of the officer across the desk, enjoying a flattered undergraduate's understandable feelings of superiority. The projected career path was being vividly described, Barnard imagined, because the interviewer's own had been interrupted in favor of campus recruiting. Regardless, that midwinter day, with its coastal chill and unsteady wind, was a fitting occasion for redirection. Years later, when himself an officer, Barnard occasionally would recall that interview and speculate on the commander's fate. It was quite likely, he concluded, that the man did complete a successful career, one approximating that which he had projected for the promising undergraduate.

His mother came up for graduation and Barnard thereafter spent only a few days at home before returning to his apartment to devote the remainder of the early summer to getting into shape. He ran more and sailed less. He felt he could easily overcome any upcoming physical and intellectual hurdles. Neither was to be the case; he had not factored in the stress, the urgency, novelty, and range of the demands that were to be placed upon him. If he had taken the recruiter's concluding hints as forewarning instead of compensatory hyperbole, he might not have made the attempt.

He settled for the superficial in many ways during those hectic weeks. It was, his intuition instructed him, an opportunity for one last fling. He would call or find someone, take a long ride down Pacific Coast Highway to a crab shack at the marina in Oxnard, his hand exploring her on the way back. Once, feeling unease that he interpreted as anticipation, his apartment still a half-hour north, he pulled across the highway onto an overlook for a hummer. It was moonless, very dark, and they were within earshot of the waves. Fatigued, preoccupied, he proved to be half-hearted. She fondled his smooth, pink-rimmed glans. Eventually her gentle bites and warm kisses had the effect she expected and he so earnestly had begun to desire.

"Where's its pullover?" she lifted her head to ask. "It should have, like, a little turtleneck thing, before. Are you ...? Hmm. Hmmmm." No one had remarked on it in that context before.

"Nope, no Hebrews in my family. Far from it," he laughed when they were again speeding north. "It's a male health thing," he explained.

Barnard's father had him circumcised at the hospital. Grandfather Adolph was livid when told, vowed that the transgression would not be repeated on any future Cordners. It was to be some years – seven months after the passing of his father to be exact – before that issue would be revisited. When watching the infant Anders cavort in his bath, the adolescent Barnard was unsure whether he preferred his own style of nether trim or not. Once an adult, he ceased to be concerned about it.

Hair agitated by the wind as they drove up the coast, she asked, "I wonder if it will feel different, being so smooth, I mean?"

Silent, he placed her hand on his crotch, squeezing her fingers beneath his so she might begin to know. At his apartment they stood in the middle of the dark room and kissed. Her chest pressed hard upon his, he could feel her resilient contours. He pulled away, unbuttoned her thin blouse then let it drop. He ran his hands over the warm skin of her back then placed them on her shoulders and slowly turned her about to unfasten her bra. He pushed its straps over her shoulders, over her hands and onto his, letting it dangle briefly before letting go. Her breasts felt substantial in his cupped hands. Weighing them, he pressed his palms against their nipples, rubbing in tight circles and feeling them stiffen from his touch. He lightly pinched the firm projections and sensed his own expansion. Kissing her ear, his hands slid slowly to her hips.

She bent forward ever so slightly and rotated against him. Her nails made a low hum as they scratched along his thighs then explored between them, measuring him. The zipper at her side was tight. The slow crunch of the yielding slider excited him even more. His hands slipped easily under the elastic across her belly, moving slowly from navel to hips then down and beyond, his ring fingers tracing the grooves of her groin. He paused at the edge of coarse hair, his lips puttering along her neck. Gratified to find her moist, he lingered for a moment then moved on.

She turned, so they could kiss, and stepped out of her crumpled jeans. He followed suit, easing off her underwear and dropping them atop his own. His hands caressed her back and buttocks, which he separated and kneaded with wide spread fingers. Round and smooth, they seemed like oversized, uncomplicated breasts. His thumbs searched for the vague hollows above, so delicious when seen. She pressed her belly against his then crooked one leg around him while he toyed with her from behind, for which she hummed her appreciation, her desire. He thrust into her, as deeply as he could manage.

Later, across the bed, he was welcomed again. Her forearms, languidly encircling his neck, tightened when he penetrated deeper. For a long moment he stayed motionless, his eyes closed, the hard member that was the extent of his world engulfed by her compressing warmth. Then he pumped, slowly, pressing his arms down firmly on either side of her head. She responded with slow upward thrusts of her hips and an arched back. They moved in equal measure, in slow primal rhythm. She tightened her embrace and pulled him in. Cheek upon cheek, she squeezed him tight, tighter.

"Mnnh. Mnhhh," was all she said.

There was a farewell beach party, of sorts, in general recognition of a last casual summer and not specifically for Barnard's early departure. But his career choice was unique, utterly foreign to his schoolmates. He had waited for the threatened drag and dump into the surf, eventually eliciting it with broad hints and a half-hearted struggle. It was a perfunctory performance by stand-ins. Dave and Hank were already off to their separate graduate programs on the east coast. Only acquaintances were left.

Self-centered dalliances and insertions thereafter gave way to pensive divestiture. Finally, he packed what remained in a pair of bags and headed east, to a southerly destination, as opposed to the cold Northeast of his two gone but never forgotten undergraduate companions. He regretted leaving the SoCal coast at the peak of summer.

Within his first few hours at the base, before Indoctrination Week officially began, Barnard sensed the squirming of second thoughts. He had cast himself into a nest of hyperactive, screaming crazies! A second tier graduate school, with a forever subordinate career to follow, suddenly seemed preferable. All he was going to get out of this experience, when he dropped out as he knew he would, was the humiliation of having to make another change in direction. Akin to the commander's advisements weeks prior, what he judged to be extreme posturing was in fact mild ritual. Via the bellowing and the barking of trivial demands by the candi-o's – the Candidate Officers – on that arrival morning and, in fact, throughout the week, he and the gaggle of soon-to-be classmates, whom Barnard has mostly forgotten as individuals, were receiving the gentlest of indications of what was to come. By their twelfth week, when they were themselves Candidate Officers, they came to recognize that this had been a well-intended, preparatory introduction that they had good cause to emulate.

Indoctrination Week was intense. The information – gouge was the new word for it – that accompanied his OCS acceptance letter was a pale forewarning of the sheer mass of service-specific material that was his obligation to absorb and to recite upon command.

Naval Rank Order! Chain of Command! Articles of Code of Conduct!

Slow answer, mistake, or blank brain was anathema. General answers were cast aside with full volume rants that disparaged unworthy ignorance, ineptitude, and sloth. The constantly shouted demands were for answers that were "by the book," their gouge books, that is: relevant, exact, immediate.

Neither was he prepared for the barrage of trivialities that defined their constricted world. He was disoriented by the jumble of inflexible rules, which were sacred solely because, it appeared, they were stated: When, where, and how to stand; precisely where his fork, knife, drink glass, and plate were to be on the table during chow; squared meals, with his movements restricted to the orthogonal repertoire of up or down, in or out. The gig line – his belt buckle, shirt buttons, and fly aligned just so – was sacred. The military terminology, acronyms as well as slang, comprised a new language that they all struggled to acquire. There was the fatigue of constant activity, constant interruption, constant impatient demands and corrections that were not for welcoming or for tradition's sake. It was not freshman camp or fraternity hazing. It was deconstruction leading to insertion and integration. It did not matter if one's undergraduate degree came from UCSB or Yale, from Podunk Tech or MIT, nor even if one were a career sailor without a pedigree of any sort. All were presumably qualified. Whether snaky geek, experienced smart-ass, deliberate plodder, or fast thinking frat boy, the rigor was the same. A great tear down was to prepare the way for intuitive unanimity built upon a common foundation.

4 - - WHO IS BARNARD?

The drill instructors were Marines. Fit and focused, hard and proud professionals, they were incapable of taking any shit, as if any sane Indoctrination Candidate would be fool enough to test them. Until the last days of their 14 plus weeks of training, the DIs were unknowable as men. The right way was easy to define. It was what they and the gouge books said it was. Barnard strove to adapt. His determination was duly but not demonstrably noted. The early advice to "Save your energy for doing, not thinking" had seemed overtly simplistic when offered, but it was the best.

"DISCIPLINE: The instant willing obedience to orders, respect for authority, and self-reliance."

It was printed right there, in his gouge book, a prime lesson for him to learn.

The yelling left his ears ringing. His own ballistic replies wore his throat to an eventual hoarse rasp. Barnard was an It, a Body, always in the third person.

"Aye! Sir!"

"Sir! Indoctrination Candidate Cordner, Indoctrination Class zero eight six eight, does not know! Sir! He will find out! Sir!"

"Attention on Deck!"

"Door Body! Stand fast!"

"Aye, section leader!"

"Shoulders tight and back straight. Understan' me? An' stop bobblin' like yar playing with one of yar sorority toys! Is this too hard f'ya, Cordner? Up! UP! Straighten UP!"

Barnard had not adequately assumed the correct brace, the proper stance of attention when the DI came on deck. His heels were, therefore, against one four-inch box placed on the deck tight to the bulkhead and his back was hard against another. He strained to hold the latter fast, the sweat welling. The DI's ruddy stubble was close. Barnard could feel the impact of his words, his breath's moist heat.

"Get that friggin' head over yar shoulders! Let tha' friggin' jewel box drop an' ya'll be here 'til FRIGGIN' lights out!"

"Sir! Aye, Sir!"

Another time, coming out of the head, he was caught off guard.

"Where th'hell you been, Cordner? You call that dressed?"

Barnard unwisely motioned to his fly.

"I don't GIVE a rat's ass!!! Zip it up in THERE, butt hole, not out here! And straighten UP! God DAMNIT. I want your head hard to that overhead, slouchbag! Your BIG goddamn head! You hear me?"

Barnard stiffened. How can he make himself taller? How can muscles, which are built to contract, do that? Barnard serially considered his tongue then, erroneously but given the DI's wording, his penis. Eyes focused on the bulkhead opposite, he started to smile then forced his lips into a tight line.

"Sir! Yes, Sir!" he yelled back. He pressed down hard on the deck, tightened further and, surprising himself, stretched his frame upward.

Militarization – reconstruction according to its own blueprint – was a central part of the service's plan. It was not by lame habit or perversity that it was a Lewis Carroll world of made up words and disconnected meanings, a world where what was right was what you were told, even if it were wrong. There was no chart to guide him other than the gouge book. To navigate, one had to accept the basic premise of discipline for its own sake. The aim was to have them, the Indoctrination Candidates striving to become Officer Candidates then Junior Officers, be constantly mindful of the real issue accompanying their entry into OCS: Were they where they desired to be? "Should be" was not the issue. Neither was "going to be." They had to want it enough to sweat and suffer cramping exhaustion, want it enough to accept being belittled and harangued, want it enough to spend hours going over and over and over what they had already polished, cleaned, and neatly arranged before the havoc of a drill instructor's stormy dissatisfaction.

Barnard was almost held back, rolled, for fouling up on his first real inspection. One additional deduction that day, for a piece of lint or a violated gig line, and it would have been off to Holding Company until he could join a later class, to continue on three or four weeks behind. That RLP – Room Locker Personnel – inspection was a huge hurdle. Another, a fast few weeks later, provided a long remembered lesson.

Each IC had to pass as an individual. In addition, his class had to pass in the aggregate in order to secure. Common as well as personal areas had to be maintained according to regs. All demands had to be met by all. Not to secure meant no liberty, withheld privileges, more frequent inspections. Be burdened with one dumb-ass or maladjusted rebel meant their entire class would be penalized. Two weeks in, Moss was easily labeled as such. In a practical demonstration of the precedence of the unit over the individual, their first bonding as a group, they formed a plan and executed it. They took turns coaching Moss, encouraging him, forcing him to get it right, to pay tight attention in class, to brace properly, to execute briskly upon command. A week can be a long time when your day begins before dawn and every minute is pressurized. The many gigs and the chits Moss had to work off attested to his lack of improvement. There was a way to ensure that their class would secure as a unit, however. At Monday's major RLP the reprobate could be shy one pair of pants, its hanger slid off to one side and empty. His locker could be in mild disarray. H-company then would have an additional IC that evening. The potential solution reminded Barnard of his FYB days in chem and physics labs, when all had been intent on garnering that top grade, many willing to deny cooperation or even to resort to sabotage for an enhanced shot at graduate school.

There was another way, a more demanding way.

Without outside prompting, the class doubled up on the inadequate IC. A seemingly organic coalescence of group intent kept him at it. They went over his bed and space, his locker and kit. They grilled and inspected, repeated the process, then did it all again. Under their tag team direction Moss polished, prepared, and practiced until nearly dawn before the critical RLP. They had him sleep on the floor after making a meticulous rack. They were gigged a few points at inspection, but they passed. The class secured. What they had done was out of reg. Their DI knew, but coming together, building unity from top to bottom, focusing on the primacy of the team, these critical aspects of leadership can never come too early. Barnard did not just witness, he experienced the validity of teamwork. He had illustrated for him the truth of the paradox that the weakness of one should strengthen the group not diminish it.

Each week was filled with stress and the pressure of urgency. Their gouge books were always at the ready – in a pocket or stuffed in a sock, depending on their state of dress – until committed to memory. His early summer exertions had been wise. There were seemingly endless runs and endless pushups, often for ad hoc infractions. "... forty-eight, forty-nine, fifty! Sir!" Drill, swim. Drill, run, drill. Test, test, test. Survival floating – drown proofing, some called it – and RLP inspections. Military Training inspections. Barnard's leaps over each successive hurdle enhanced his preparedness for the next. The classroom material was not difficult, only there was so much of it, with never enough time. The sensory overload and time pressure were deliberate. Only a planner bound to a desk has the luxury of organizing and reviewing so as to suffer no error. Military decisions must be made quickly, often with incomplete knowledge of the facts or even of what might be additional relevant factors. Officers must cope confidently and project that confidence onto others. Civilian executives make decisions; service officers must often make decisions under extreme stress. Whoever cannot do that is not useful, cannot be relied upon by the group, is not wanted. Accepting the logic of it made the continual stress less of a burden for Barnard.

There was no ease up when they became candi-o's. In fact, the pace increased dramatically. By then accustomed to precise discipline, having absorbed the applicability of their training, having acquired a degree of inner as well as outer tempering, they were less dismayed by the more intense schedule. There were classes in damage control, sea law, fire fighting, warfare, seamanship, naval history. Repeatedly lifting their chins into a thin pocket of air as they worked, they learned to remain purposeful as a simulated line break brought cabin water level to within inches of the overhead. Potentialities begot trials; experience was gained and skills were acquired.

Barnard showed a fair degree of aptitude in his side arm quals, which surprised him. He had not handled a weapon before. The controlled tension of firing a pistol accurately, the gun's recoil, the impact of its blast upon his chest, even the odors, of gun oil and burnt gunpowder, were not unpleasant. It was another unanticipated hurdle surmounted.

OCS left a largely positive impression on Barnard. He regretted none of it. It was emotionally satisfying to brace with the others, to hold heels and head precisely so and shout "Hooyah." Humiliation and fatigue faded, leaving pride. There were good days, days of release, weekend liberties. At last came the victory run, the commissioning oath, and that well-earned salute. Many had wives, girlfriends, or family to watch the newly christened Junior Officers receive their gold-plated Ensign bars. Barnard had no one. He regretted that Mother did not fly out for the ceremony.

He had signed on for undersea nuclear, the Submarine Service track, without any real recognition of why. His choice may have been influenced by hints that the food was better and the formalities more relaxed undersea. It could have been because he preferred being part of a compact rather than a large, diverse crew. A submarine carried a complement of approximately one hundred – half again more if it were a guided missile platform, a boomer. His preference might have been an externally derived insertion, one encouraged because his suitable physique rather than an innate desire to be one of the select who would wear the Golden Dolphins. That symbol of achievement is still in a velour-covered box, somewhere amongst his underwear and accessories. Often taken out to hold for long, silent moments, he had planned from the outset that it would someday go to the son he knew was in his future.

Upon becoming a Junior Officer, Ensign Cordner began a full year of Naval Submarine School, half of it in class, the remainder at a land-based nuclear facility. Even though it emphasized application over theory, he was glad to get back to academics. It was not the comedown he had sometimes fretted over. It was not learning to be a glorified technician and he did not take offense at being termed a nuclear engineer. Intense science could come later, if he chose the path of further graduate work after completion of his initial service obligation. He did well, finishing up as third in his class.

At Officer School, he learned of the Navy's ways in more detail. He was tasked with acquiring the set of skills that would make him a reliable component of the team. Barnard felt he lacked an intuitive feel for Leadership with the capital L. He anticipated being hesitant when faced with such a presumption. Through the deliberate molding, he nonetheless became comfortable with the manner and necessities of command. It was a fast-moving three months.

When he reported for his first tour, he stood on the pier to study the moored, tubular craft. It was a nuclear missile sub, an SSBN – what he had hoped for but had known was pointless to request. The boomers were roomier than the fast attack subs. Motionless, low in the water like an abandoned piscine toy, the boat was sleek-skinned and dark, looking much larger than its silhouette in Jane's had suggested. Its clean hull showed few breaks of contour. Right, he mused as he scanned it aft to fore, an immense, grim toy. The figure standing atop the bridge in the sail, looking laughably tiny from Barnard's perspective, reinforced this impression. When the crewman disappeared almost as quickly as he had popped up, he appreciated that he was about to join an experienced cadre of officers and sailors. Sensing his minimal competence, he hoped that he had drawn a tolerable skipper.

Busy, dim, green, inducing vague claustrophobic anxiety – these were his major impressions inside the boat, and not totally unexpected. The same color was used throughout, with a superimposed riot of signs, notations, and symbols. He had toured an undersea boat long before, then experiencing the craft through the eyes of a visitor. This time, from sub school, he knew what nearly everything was, even if it required a few seconds to get it exactly right. This partial familiarity was welcome. Nonetheless, it was several weeks before he adapted to the complexity of this particular craft, to the many congested passageways and watertight doors, to the cacophony of sharp klaxons, ragged intercom squawks, and alerting tones. He had to learn to be constantly alert for head-cracking projections from bulkheads and equipment.

As an officer, Barnard could eat in the wardroom. He was not restricted to the general mess. Personal and sleeping space was another matter. There was not enough of either for all to have assigned racks, so the seamen rotated in accordance with duty schedules. Until he earned his Lieutenant's rank, Barnard, along with the other Ensign, who had reported a few hours after him, would be sharing space with the crew. It was part of submarine protocol and not as onerous as he had feared; profound fatigue made sleep easy, wherever and whenever. Still, Barnard looked forward to getting his qualifications signed off, moving up in rank, and earning his own bunk.

There were no external clues to disturb the eighteen-hour activity cycle, the schedule that the Navy had determined made most efficient use of men, time, and space. It comprised six hours on duty then two six hour periods for dealing with secondary tasks: rack time, studying, eating, personal care, and rec. Not much rec. Barnard's bodily rhythms quickly adjusted. He got used to short showers, fast meals, snappy pops out of his rack. He developed the knack of awakening promptly and voiding quickly. There were satisfying feelings of accomplishment, of personal progress as the weeks of patrol went by.

It was remarkably smooth sailing when submerged, no matter what the weather. And the food was fine, even better than he had expected. Barnard had to curb his appetite, use the fitness equipment regularly to keep his weight down. Other features of the boat were notably less pleasant: the minimal privacy, the constant noise, the fatigue of repetitious drills, the constricted space and loss of visual perspective. These impressions held fast, shaped his later preference for sail boats and his disinclination to be confined.

Carpenter, the other Ensign who came aboard with Barnard, was one OCS class before him. The small difference, less than a month, made Carpenter the Bull Ensign, which set apart the two minimally useful virgins. Given a range of distasteful duties, the kind of unpleasant, often thankless, necessary tasks that fell upon junior officers, Barnard always drew the least desirable. It was part of the coming together, of learning to lead by being willing to follow. The need for instruction and repeated practice tempered the impositions endured by the newbies. Patrol was not another indoctrination; it was for real.

For example, on solid, stable ground at sub school, he had been taught the theory and basics of sextant navigation. At sea, the task was notably different. Standing on the slippery decking of the congested bridge, tensing to steady his arms, he learned anew. He would point the instrument at sun and horizon, determine their relative angle. At night, under a partial moon that emphasized that the boat was the full extent of their solid world, he would carefully hoist himself and it up through the dank sail to similarly shoot Polaris from the same high perch.

Chiefs, the senior enlisted seamen, each with specific expertise, were a leveling influence. They were the primary link between the JOs and the skill set that had to be acquired before they would be recognized as having even a hint of utility. A book or class was not enough. Under their generally patient eyes, the subtle complexities of damage control, plotting, diving, propulsion, and sonar were imparted and mastered. It was hands-on learning, a form of tutelage with focus, diligence, and honed instinct favored over impressive smarts. Barnard can remember lying exhausted in his rack, sing-song reviewing the range of personalities he was working with:

"Good chief, bad chief, zealous chief, prick."

Barnard affected a haphazard beard during that patrol. He liked the mature look of it, but not the thing itself. It caught bits of food, it itched, its shape and trim were uneven. Nonetheless, it meshed with his image of a submariner. It was to be his talisman for bonding with the experienced crew. The other officers, even the captain, would explore its configuration when they spoke. With the Bay of Pigs in not too distant memory, a few of the enlisted complained that he resembled Castro. He would occasionally mouth an unlit cigar to complete the unfavorable comparison and make light of the crew's displeasure.

Barnard's service started while the Cold War was still hot. The mental set was of Us versus Them; white hats versus black hats. Frying the villainous commies, if it came to that, would have been done with practiced proficiency. During their three months on patrol, undersea for much of the time, the crew was constantly ready to launch nuclear missiles according to pre-established targeting criteria. This was their function, their purpose as a unit of highly trained individuals. There was no point assaying the morality. It was their assignment, their duty. Barnard, along with each of the others, had to be able and ready. There were unending drills to solidify this: incendiary drills, using fans rippling red and white cloth to simulate fire; leak drills made realistic with sea water; equipment damage drills, often with ancillary pieces purposely put out of action by the drill coordinator; casualty drills, with ersatz blood and simulated injuries. There were lost man drills on those occasional days on the surface. The exercises did not always originate within the boat. Even the captain, Commander Grennyman, could not be immediately sure whether a received action message was drill or factual. The prelaunch sequence would proceed with practiced precision, right up to the last steps of targeting the missiles and making them ready to arm. The codes, the multiple layers of authorization in place at that time would seem quaint to any modern sailor, even an ordinary. Barnard's then circumscribed world was only tentatively entering the digital age.

During his first active participation in a launch exercise, Barnard felt his shirt clinging to his back and glanced sheepishly at nearby uniforms to gauge whether he alone was so stressed. He was not. Not all of the crew were deeply experienced sailors, most were just more experienced than he. Sweat, bruises, anxiety, and fatigue were pervasive. These eased as Barnard and they acquired the skills that incrementally increased their utility.

"Timed course correction," he heard after the predetermined interval since the last had past.

Tense from the stress of being unerringly precise, Barnard saw a droplet from his chin hit the plotting table.

"Aye, Sir. I, uh ..."

"Decide first, Ensign. Then speak."

Ensign Cordner straightened then gave the proper heading, his voice as firm as he could make it. There was no request for confirmation, no inquiry as to whether he was certain. The course adjustment was passed on and repeated back. The boat slowly responded to port.

There were no fires, hull damage, breakdowns, or casualties. No one fell off a slippery deck into churning waters. No missile was sent on its destructive flight. As the patrol progressed there were intimations of conflict. They trained as the rabbit, sometimes the wolf. They never saw action in any true sense of the phrase. The occasional contact or incident was duly logged and reported. But the officers' and crew's judgment that they were ready was not enough. They had to prove it again and again. They had to show, in concocted situations of every sort, that they could be relied upon to act as a team even to, especially to the unexpected. All were constantly tested, with the aim of expanding their repertoire and thereby assuring smooth execution of the extrapolations necessary during actual combat. The tight complement serving on a submarine can feel as one that growth of confidence that accompanies the acquisition and proof of adaptive flexibility. Big companies, states, and nations have the same need, except they span too great a range of attitudes and goals to easily sustain a comparable sense of unity. Often they try to make do with artificial, often childish group activities drawn from populist handbooks. When Barnard later looked back upon his naval experience, he would reflect on how their training had been far beyond that, far more rooted in their reality.

The enforced familiarity of close quarters, the frequent, frenetic and always timed training drills, the stress of the confinement away from normal life, each contributed to the pressure that would build for the crew until their patrol was completed. At the halfway point, five to six weeks in, it was traditional to have a blowoff party. It was something to look forward to, something that blunted, for a while at least, the dull annoyance that accompanied the realization that an equally long period of isolation and hectic activity was ahead of them.

They surfaced in the mid-Atlantic, well off the shipping lanes, under a bright, featureless sky. The deck, the steel beach for their celebration, was covered with off duty bodies absorbing the sun and sniffing charcoal smoke. This was Commander Grennyman's sign off patrol. He was, with mixed feelings, moving on to an instructor's slot. It was the correct career step for him, potentially the prelude to that Senior Staff posting in the Pacific that he had been striving for. For his final halfway he had stashed additional prime filets and huge lobster tails in the locked freezer. These, cooked in stages so the entire crew could partake, were being enjoyed a few feet above sea level.

The word was passed for the officers and chiefs to gather in the wardroom, as loosened duty schedules allowed, for an urgent meeting. When Commander Grennyman entered, he looked uncharacteristically subdued. He sat heavily in the arm chair at the head of the table then stared over the heads of the assembly with a perturbed frown that was far from his usually attentive yet placid command bearing. The scattered conversations gradually ceased.

What the hell was going on?
Had some dramatic action directive come in?
Was this next burst of activity going to be for real?
The questions careening through Barnard's brain were, he imagined, being shared by all. He looked across the table and raised his

eyebrows. The Exec offered a blank expression back. If something that serious had happened, then the entire crew, they in the wardroom, and certainly the Exec should know so that they could begin to prepare. There was a knock. McMatley, the Chief of the Boat, entered. Grennyman looked at him without speaking for several heart beats then reached into his pocket and brought out three steel balls the size of marbles. They clicked and clacked as he manipulated them. He stared at McMatley. One of the officers in the tight space let his head drop, which Barnard did not notice.

"Did you get an answer for me, Cob?" the captain asked.

McMatley stared back, meeting Grennyman's dour gaze with his own. The anxiety seemed thick, like an enveloping fog.

"Sir. No, Sir. I did not, Sir."

Those infernal ball bearings, the sound of their metallic collisions dulled by the captain's cupped hand, dominated the hushed officers. Grennyman shifted in his chair, looking sideways then down, working his jaw.

"Then report status, Cob, damn it!"

The room itself seemed to hold its breath.

"Aye, Sir!" McMatley barked smartly. Then, after another short pause, he stated crisply, "Captain, there's not one damn strawberry to be found. The beard must've got'em."

Barnard felt his heart stop.

"Then ... we'll have to make do with ice cream," Grennyman said slowly with the start of a grin.

Click, clack, click. Then came a roar of release. They had been had. Barnard, the junior of juniors, had been had most of all. Suddenly able to think of something beside the present moment, he did not need the origin of the brief skit spelled out for him. *The Caine Mutiny* film was relatively recent and, even for the officers, movies were not the typical mode of entertainment while in port. Only a few, Barnard being one, had seen Bogart click-clacking before the court-martial in a similar sideways posture of disconnection. The laughter went on for several minutes, those who knew slapping the arms of those who had to be enlightened.

Commander Grennyman navigated the room, speaking casually with each of his officers. He had wanted to start his farewells then, at halfway, with laughter in his ears, instead of at the end of the patrol when he knew making a lighthearted departure would be difficult. Barnard could sense nothing truly specific in his comments to him. He did appreciate the gesture of a pat on his shoulder and the look that indicated that the joke at his expense had been because he was the one at the bottom of the officer ladder, no more than that.

"Shave it off," Grennyman said finally, pinching Barnard's chin and conning the hairy face from port to starboard and back. "It doesn't work for you."

To be the focus of the put-on was a mixed experience Barnard would later recall as virtually singular. As silly, as stagy as he knew it to be, the slight drama, born of upcoming separation and truly aimed elsewhere than at Ensign Cordner, nevertheless made him less a stranger, less a newbie. Without any notice of the point at which it might have occurred, he was one of the crew. That was more rewarding than the overwrought skits, raunch, home movies, lazy sunning, and carefully monitored swimming that would come later that day.

Barnard does not remember the manner of Commander Grennyman leaving the boat for the last time. He barely remembers any of the specifics of his short-lived naval career. They exist as disjointed fragments. Sadly, one does not always remember the things that were meant never to be forgotten. What he has retained is an appreciation of the power implicit in the fact of a diverse group coming to share a common goal. There were the nukers, the pingers, the peepers, and the engineers. There were the torpedo men and the reactor men. Each saw himself as critical, yet each knew that any presumed central role was an illusion. The entirety was what mattered; each served, in his special way, the purpose of the whole.

In this respect Barnard felt a degree of unease. His future was for him alone to decide. He recognized his capabilities, his intelligence. The question was whether he would remain fully committed and be secure under pressure. In his rack, savoring a few hazy moments before falling into a welcomed oblivion, Barnard revisited what the Exec had told him after his most recent qual review.

"The Commander prefers, and you'll find that most of the captains prefer that an officer do everything equally well, not just some things outstandingly well." He was praising the utility of patient consistency, reinforcing the practical necessity that, at critical moments, a negative is not cancelled out even by a plethora of positives. "You tend to think a lot, Cordner," he had gone on. "That doesn't always help. Save it for when you're stationed at a desk."

Unfortunately, it still was hard for Barnard to rely on instinct, even if finely honed. It seemed a frequently manipulated internal, one he did not always trust. He preferred the relative security of certainty before acting, which added critical seconds. He feared that this latency would do him in, potentially do them in. On his next liberty, when he was again a pale explorer of museums, shops, bookstores, and historic sites, for a few days free of the constant pressure, his doubts blossomed. Barnard began to seriously question whether he was truly on the right track.

It took several patrols for him to get the final necessary chits in his qualifications book signed off: reactor, countermeasures, communications, launch procedures, missile status. As his skill set progressively grew, he moved up the ladder of responsibility: Engineering Officer, Duty and Watch; Contact Coordinator; Officer of the Deck. His final grilling before becoming Qualified in Submarines was intense. He was calmed by the fact that he would not have been up before the grimly formal officers if he were not deemed ready. None of the examining board had forgotten the USS *Thresher,* the first nuclear submarine to be lost at sea. Barnard was barely aware. The *Scorpion* tragedy was relatively recent. Barnard was well aware. Some lapse or error, some overlooked contingency or improperly managed failure, some unanticipated flaw of men or machine had sunk her. Neither event was openly discussed but each had lessons to teach. The questions put to him, therefore, were confrontational, detailed, and conflicting, as his duties would be when real emergencies arose.

They were underway when he qualified. Therefore, sharing the event with the other officers had to suffice. He was proud to have earned those Dolphins. The insignia was the tangible symbol that a significant level of accomplishment had been achieved and recognized. Years later, his doctoral qualifying orals and dissertation defense seemed like "child's play" in comparison.

"In my opinion, you're excessively deliberate, too tied to your emotions. It's going to get you into difficulty. And, if I can sense it," McMatley said privately a few days later, midway up the sail, "others, the crew, will also. It's something you need to work on."

It was a lousy time to be getting that sort of advice, albeit informally. If they had their doubts, Barnard brooded, they should not have qualified him. The arrival of several replacement officers, when they were next in port, provided Barnard additional opportunities for reflection.

"The Navy isn't interested in flashes of brilliance. We, the crew as a team, can't rely on that," the new Exec said to him over coffee, while waiting for the ordinary seamen to come aboard. The comment came as a frank observation and unrelated to any direct question from Barnard. "We need steady performance at a high level, not peaks and valleys." After a long stare into his cup, the Exec looked up and continued with, "If an officer's not one hundred percent committed, then someone else has to pick up the slack. It's the same for the ordinaries. It's a job and a half here."

It was information, gouge not criticism, something appropriate for him to impart to those of lower rank. Everyone was continually being evaluated. If an officer or crew member possessed a negative trait or even a less than intense commitment, he seemed to be suggesting, then it was best to have the flaw explicitly noted and addressed.

"We'll make the decision if someone's best is sufficient. Yours clearly is, from what I've reviewed," the Exec then added. "It's up to you to decide. Sufficient may not be enough for you."

Barnard was relieved to hear the positive assessment but surprised by the final comment. He should not have been. Fitness Reports are not just passed up the chain of command. They are living documents, carefully compiled and reviewed, and intended to be exactly what they are called. In any event, the Exec's comments touched a nerve. Throughout the five years Barnard had logged in naval training and service, secure and sincere commitment had been elusive. For close to a decade – at UC Santa Barbara, then at Naval OCS, Naval Nuclear Power School, even into much of his first tour – for nearly a third of his life, what he had received far exceeded what he had given. The key criterion hereafter was, indeed, going to be performance, not potential.

What if he could not deliver?

If he did leave the Navy, it would need to be with a solid plan. He was too old to chance another false start.

Barnard's self-doubt triumphed and he decided to resign his commission to pursue a graduate degree in something other than physics. Although it would have been feasible to pick an Ivy after signing off, the Pacific Coast was the only coast for him. He set a course for Claremont Graduate University, where he would study economics, intending to focus on the quantitative side of finance. On breaks between patrols, he assembled the application materials.

Yale, in New Haven, was an hour or so drive from New London. He signed up to retake the GRE there and, luckily, his patrol schedule did not change. Despite that it seemed he had been away a long time, the co-op there revived pleasant memories. The girls were attractive in their designer jeans and, it seemed, totally independent. The guys appeared immature by comparison; Barnard was mildly annoyed by their selective attention to a circumscribed world. His one night stay was too focused to take part in the collegiate dedication to personal success and the sensual life, too brief to dip even his toes into the font of consumerism that was being engineered to replace the often contradictory posing engendered by the soon to end Vietnam War. Walking the campus after the test, he again perused the numerous posters and announcements, which seemed to focus on the immediate, on the comparatively unimportant or, worse, on the marginal elevated by special interest to the level of urgency.

His test scores revealed that he was not significantly smarter for his naval experience. Still, they were more than adequate. He proceeded in earnest. There were letters of reference to ask for and to have sent. His application packet almost missed the deadline because Barnard initially overlooked that his undergraduate transcripts needed to be official. He got it off in time, the echo of the Exec's words still in his ears.

Back aboard the sub, he reviewed the engagement tactic they had practiced while deep in the Atlantic. If the sonarman detected the whine of a closing threat, evasive maneuvering was the sole option. Before that, if they knew the boat were being hunted for real or even for training/sport, an established procedure was for the rabbit to go to All Stop and wait, thereby to be a passive, silent, uninteresting, subsurface lump. Torpedo guidance systems were designed to ignore anything seemingly stationary, which would most likely be an undersea mount or hulk of a wreck and, therefore, a false target. Remaining as you are, in other words, can often be the best course. It is usually the easiest. Barnard weighed his options repeatedly. He reread the acceptance letter many times before making the irrevocable decision that would have him salute the flag for the last time.

Located at the eastern edge of Los Angeles County, Claremont was an hour drive from the beaches and the marinas of Newport, if the freeway was not snarled. Barnard appreciated not having those additional temptations made too easy; he would need time to adjust to his new life.

Training and experience work in strange ways. What one expects to find difficult often comes easily. The reverse is common as well. Upon becoming a civilian, Barnard was no longer in a world dominated by arcane customs and rigid regulations. He had full freedom of action, with an overabundance of choices. This exposed that there was a part of him that welcomed being firmly directed. Compounding his disquiet, he found himself dwelling at length upon what he had given up and reviewing alternative futures that would never be tested. The choice has been made, however. Reconsiderations did not yield regret, only new uncertainties.

His apartment, not far from the campus, seemed huge. He had an extravagance of personal space. He could stand in the bedroom with arms outstretched and spin without hindrance. He could shower for as long as he wished, feeling the twin luxuries of unending hot water and an absence of urgency. The shades on his living room's multi-paned windows were always open. Day or night, the outside world was just a glance away. But he had sea ears in addition to sea legs; the apartment was immersed in unfamiliar sounds, which accentuated what he missed. There were no keening gulls, as there was no shoreline. There were no waves, so no dull thumps and prolonged wet rustling of their collapse. There were no heavy footsteps on metal, no raucous alerts from horns and klaxons, no steady low rumble, no gentle rolling. He was slow to adjust.

At the college, Barnard's was assigned a number of make-up courses, most of them soft topics far removed from mathematical physics: Macro and Microeconomics; Public Economics and Enterprise Economics; Empirical Statistics, which focused on applications. All were easily mastered. Then came courses on the Firm, Sovereign States, Taxation, Game and Decision Theory, Federal Banking, Economics Across Borders, Modern American Capitalism, and Corporate Finance.

The jumble began to take shape as Barnard closed in on his initial inclination toward Econometrics, detailed quantitative economics. As preliminary as was his background in mathematical analysis for a physics major, it seemed an advantageous head start for an academic career in economics. It was not until decades later, however, in the nineties, that quants would be the sought after rock stars of finance. Barnard was too early to catch that wave. Computers were still large, exotic beasts used for laborious tabulation, not the ubiquitous foci for divination that they would become. Manufacturing, markets, and money – key, then as now – were integrated and governed by the principle of maximal utility. Quantitative attention to transactional detail was useful for analytics, not prediction. The placid, scholastic atmosphere was a pleasant change. It suited Barnard at this stage, which was quite appropriate for one unlikely to achieve fame and virtually certain to avoid fortune.

As he relaxed into civilian life, he merged with the campus, with the activities it offered. Finding the dark confinement of movie houses too insular, he acquired an interest in live music and theater. Nearby Scripps offered other pleasantries. He enjoyed meaningless encounters with pretty, touchingly earnest young seniors. Sliding his hand over silky fabrics and skin untouched by the sun, kissing protuberant nipples, and exploring moist crevices were realities much improved over his recent past. He craved contact and connection more than physical release. He wanted to be touched, to have erased his fantasies of untested port whores.

"Find the man in the canoe," one outspoken tall blond had said.

"What?" he asked awkwardly.

"The G-spot."

She had to guide him.

Barnard took as much advantage of that liberal interval – the Pill had arrived, AIDS had not – as his course work left time for. He also could be alone if he chose. He had regained the privilege of quiet privacy and had no urge to contemplate marriage or family. It was a treat to have rooms with walls beyond his reach, to have a bodily rhythm that matched a visible sun's cycle. He had days that were apportioned as he chose and enough in the bank for a small sailboat, soon berthed less than an hour away.

His discovery of Maurice Allais's paradox came while scanning journals for a term paper in the Psychology and Economics course being offered under Advanced Topics. This was a decidedly novel pairing at that time. How people exercise economic choice was largely the province of the marketing, sociology, or psychology folks. Few in economics cared why rats worked for pellets or why similarly captive undergraduates chose one payout over another. The scholarly explications focused on the how. For Barnard, the frankly off-center course offered easy credit hours while he prepared for his candidacy orals. Besides, the instructor was very attractive, unusually lively, also earnest and not much older than he. But she was married and so provided only remote fantasy.

The hum of the library's cool fluorescent bulbs, its characteristically redolent and tightly spaced stacks, the functional utility of its glass-enclosed study areas all reminded him of his undergraduate years. Encouraged by his instructor's enthusiastic approval of his project, Barnard spent much time there, scanning the papers on Allais's reference list, skimming over the tables of contents of the arcane journals in which they were published, often diverting to the full source of an intriguing abstract. He was like a tourist in a new land. Barnard's interest was enhanced by Allais himself, an economist who had made a parallel career in physics, albeit one with questionable relevance. It roughly paralleled his own life in reverse, Barnard imagined.

Among the basic principles taught in economics is quantitative utility, i.e., fiscal efficiency, and the undeniable logic of allocating resources so as to maximize it. Basically, Allais's thesis was a technical argument for the necessary *in*efficiency of markets. It was a critique of the "standard model," which took economic benefit to be the sole basis for decision making by the individual and, by extension, the entire consumer class. Allais's attempt at refuting that view seemed to have been received as a novel aside, as a not immediately useful insertion into the literature of economic theory. Being new to economics but familiar with physics, Barnard naively found merit in Allais's point of view, took thermodynamics as a helpful analogy. In that branch of physics, he reasoned, macroscopic phenomena, such as heat, which depends upon an enormous number of interacting and individually unobservable molecules, are indeed made tractable by looking at aggregate behavior. That is, one could incorporate what the molecules did while ignoring their individuality. Human behavior, however, is often irrational, not constrained by physical law. Allais's argument was that restricting economic choice to a calculation of general economic utility reduced complexity by ignoring individual identity, by ignoring their idiosyncrasies *and* what they actually did. De-emphasizing individual behavior in economics, for the sake of analytical simplicity, was not, therefore, a parallel to excluding the movements of identifiable discrete molecules in thermodynamics.

Barnard did not then have the depth of understanding to go beyond summarizing and restating – aka plagiarizing, in typical thematic, expository fashion – Allais's work for his term paper. Happily, his professor liked his final submission. It stood out from the predictable rehashing of ongoing experimental psychology results that were stacked on her work table. With her encouragement, it formed the basis for his doctoral dissertation: "The Implications of the Allais Paradox for the Mis-Application of Utility Theory to the Individual."

His proposal garnered a mixed reception from the classical economists on his candidacy committee.

"And what exactly, Mister Cordner, would you say does learning the psychological preferences of the individual tell us about aggregate economic choice?"

"You need to spend more time with the basic texts," was another's more direct and equally acerbic comment.

His good grades and successful dissertation defense ensured him of the prospect of at least an entry level academic post. A specific positive response to his dissertation, which he tightened up and had published, was soon to come.

His mother, Frances, from whom came his middle name, was aging but wanted to be at his hooding. He was heartened by this and arranged for a car and driver. It was a rare, pleasant visit; theirs had never been a close relationship. Standing on the platform, surrogate certificate in hand, and seeking her out in the audience, he envisioned Father among the indistinct mix of attendees. With Mother safely on her way home, Barnard toasted his doctorate with a few friends, then caroused for two days and what seemed like four nights in Las Vegas.

Months earlier Barnard had visited with the Chair of Economics at Santa Monica College, or City College as it was once called. An unexpected defection had left them with a lecture gap in their basically undergraduate courses. His advisor had put forward Barnard's name when queried and, assuming he graduated on schedule, felt certain he was qualified. Barnard was delighted to accept the post and be able to put off permanent job searching for a few years.

There were handouts and syllabi to review, a text book to annotate, and a course sequence to revise. Along with freshening up and selling his boat, he had to find a place to live. Nearby Venice appealed to him. It was not yet gentrified with tall condo blocks and developers' machinations. It smelled of salt air and was lively. Barnard found a tiny but comfortable rent house within walking distance of the marina. Except on weekends, Barnard's indulgence – a sleek Datsun 240Z – sat in an unrestricted space down the block, while he commuted via a red and white Yamaha two-stroke motorcycle that was almost as breezy as the boat that he missed but could not quite justify replacing. A year flew by before he bought a helmet, something that his soon to be so important Diane was to insist upon if they were to get serious.

His academic duties and long term plans left Barnard with little time for other pursuits. He intended to use the coming summer break for starting work on a research project, something that would enhance his curriculum vitae. He talked to a few people in the economics department at USC, for ideas and as potential collaborators. Meeting Diane, then a Psych student and part-time chorister, turned out to be the single significant result of these explorations. Her mentor had suggested that she might prove useful working with the two of them on a novel, Intentionally Misrepresented Shared Reward experiment. The inter-departmental, inter-campus collaboration did not materialize, but Barnard was smitten. He crafted an opportunity to meet up with her at Marina Del Rey, where she had a condo that overlooked the berthed 26 footer that she had casually mentioned – part of, in fact the residual bulk of her parents' estate, he later learned.

Even in bright sun, her skin was silky smooth, unwrinkled even when she laughed. She was slim, athletic, conveyed the Near East in looks and manner. Barnard was fascinated by her dense, auburn hair, which a stiff wind would make run wild if not constrained. She had dark, widely set eyes, a teasing overbite and prominent cheeks, plus a sexy shallow cleft in her chin. Most of all, there was her short philtrum, under an unreconstructed nose, which left her upper lip faintly tented and expectant, her mouth so ready to kiss. He perceived her to be beyond what he deserved.

He drifted into an easy Southern California academic lifestyle: teaching, committees, writing papers, the beach, and sailing with Diane. With remaining at SMC forever not on his agenda, he made a point of attending economics seminars at UCLA in addition to USC, in order to make the acquaintance of faculty and to drop off more copies of his recently published work. He enjoyed driving fast, sharing days and nights with Diane, and having a new sense of control over time. He was again, or arguably for the first time, a free agent. It was an illusion, no doubt, but pleasant nonetheless. His course offerings were well received, as was validated by the Outstanding Teacher plaque that appeared on the wall of his office. Otherwise, outside of Diane and a few of her musically inclined friends, he was a loner, kept mostly to himself at SMC. Some took him to be stiff, aloof. "Military," had read one of the tendentious student critiques.

Diane filled a void, one that he had not previously perceived and that he much later appreciated most profoundly upon her absence. She was the embodiment of fun. She validated sharing. She taught him tenderness and extended his boundaries. She was lyrical and rhythmic passion.

She completed her studies the next year, his second at SMC, and was debating what to do next. His memories of those days remained vivid and easily recalled for decades: Split sausages, grilled onions and peppers, all under red sauce on crusty Italian bread, with small plates of spaghetti and diminutive glasses of wine to either side, on a checkered cloth at their ristorante on Pico; the surprising shift of wind and heavy chop west of Santa Catalina while they were in the cabin and otherwise engaged; walking Santa Monica pier listening to the throaty, old-world music pacing the merry-go-round. There was an apartment up the stairs behind the carousel, they had been advised. It was available, but after a walk through they realized that novelty would soon depreciate to annoyance. A generously-sized apartment in a building with interior stairs would be the wiser choice, if they in fact agreed to make one.

They did.

He had loved her for over a year, every bit of her. They married, on a perfect August afternoon, at a nondenominational church on the edge of Hollywood. They stayed to themselves on the quick honeymoon trip to Hawaii, enjoying understandable isolation.

Nothing changed very much upon their return. On weekends they ate tangy small shrimp from foam containers after visiting art galleries in Newport. They explored still locally owned shops and independent bookstores in Westwood then drove a few miles south for grilled steakburgers "cheesy with" and fries at the Apple Pan. They braved the stiff wind atop a hill overlooking Malibu as the sun set and they nibbled chicken from a box. By the time he received that unexpected envelope with the Stanford return address, Kyle was already on the way and they were in serious negotiations on a house. He knew no one in Northern California. The letter was from someone named Kvorsky, who had been shown a copy of Barnard's doctoral dissertation and thereafter read his published journal article. The typed, signed note inquired if "Professor Cordner" would be available to give an afternoon seminar. Barnard waved the invitation at Diane and danced in manic glee as he imagined the long string of consequent possibilities.

"Palo Alto! Gee Sus Christ!" he yelled.

A serious academic was combining economics with psychology! This meant it was going to be a fertile field! They had dinner out to celebrate.

His reply listed several possible dates. If he made a favorable impression, he could be among the early pioneers.

If. Damnable If.

Barnard had decided he was unsuited for a career in theoretical physics. He had similarly questioned his naval career prospects and chose to abandon it. Late that evening, staring through the photo of Diane and him at the bow of her, now their boat, a familiar nagging uncertainty arose. There was the strong possibility that his ideas seemed so promising to him because of his own ignorance. He was expert neither in economics nor psychology. Except for that first paper, no top tier journal carried his work. Even the junior faculty at Stanford undoubtedly had years of study and application beyond his. So what if it were novel and cutting edge, a new program. He visualized being perceived as a dabbler, a finder and manipulator of the obscure. In front of an adept audience he would be a one-hit wonder at best, an easily humiliated dilettante at worst. He would need to do more than attack the Efficient Market Hypothesis. That was already being done. He would need to provide more. Compulsively, on that tortuously long night, Barnard reviewed what questions he could and could not answer, how being made to look the fool might feel. A disturbance inside, the heavy pulse and dry mouth of anxiety, signaled that the confidence that should have been instilled and nurtured while in the service had faded or never was. The Stanford opportunity was exciting, but it was also premature, he ultimately decided. Barnard rationalized that he needed to be thoroughly prepared before taking advantage of it.

In a follow-up letter, Barnard stated that he would be unable to accept the invitation. He fleshed out a tepid excuse and chose the safer, the easier path. A few coins sufficed for the stamp on the envelope that relegated him to the status of junior college professor. A singular opportunity was passed over. An imperfectly charted course was left unattempted. He had not allowed that, while a risky venture, it would have been one with enlightened company. Many trials would occupy those engaged in exploring that contra-classical subset of economics; all would need to rebut sharp critiques. He had no idea that one day Behavioral Economics would muscle aside the Efficient Market. If only he had the strength of will to claim a stake in an unclear future, the gall to take up the cloak he wished for, even if being potentially undeserving. If only he had let circumstance, not insecurity, condition failure. The sting of his regret can still plague him in the middle of a disturbed night or float forward during quiet moments of the day, like a persistent, melancholic musical phrase, a falling chromatic with decrescendo.

They moved to a house on the sunset side of Second Street, well south of Santa Monica's busy downtown. In concert, Barnard and Diane dealt with the early mysteries of birth and nurturing. There was Kyle, the son he had long anticipated, with Katie soon following. Bonding with a daughter was something he had not anticipated and was more challenging than he expected. They enjoyed their handsome boat, the near-shore sailing on easy currents, listening to the hull cut through calm water or to the rumble of light wind bothering a slack sail. When the children entered their teenage years, it was obvious that they were happier when elsewhere other than on the boat, which scuttled their parents' assumptions regarding genetic predisposition. Barnard resisted giving up the 26 footer. He relished the sense of expansive freedom that it provided during bright afternoons of occasional solo sailing. Soon, however, being aboard the medium sized craft, which they had previously handled as a couple, made him pensive. It was superseded by a much smaller Sunfish, similar to the one he had enjoyed up north years before. That, also, was eventually sold, in favor of a rebuilt, stick shift Alfa Romeo Spyder, in which he would take long, sun drenched, weekend rides with the top down. Enveloped in the Giulietta's noise and wind, he would drive fast and pretend to think. It was not the same as being alone on a boat but close. It sufficed. Life on a submarine had left Barnard with a firm dislike of being closed in, of closed windows and closed doors, of being in one place too long, of being constrained. The clash-thump-scratch of a hatch, of dogs turned to, were distant images that he preferred not to reawaken. It was not raw claustrophobia. It was the implication of limits that disturbed him. The open sea and, later, his serial open cars, were anodynes.

They loved, enjoyed, planned, and occasionally fought. Together they wove the fabric of a pleasant life. Teaching was easy. Barnard's published papers coalesced into the nucleus of a planned economics text, one with a less traditional orientation. Eventually, his focus drifted to administration. Both children went to UCLA. Kyle chose the immediate income of a mid-level software applications management position over a much-advised year or two in graduate school. He married early, again ignoring his father's advice. Katie, with her profound lack of discipline and firm sense of self-approval, wandered the fringes of tenuously justified, reflexive activism with one companion after another, none of which Barnard felt was worthy. She grew into an intense young woman estranged from parents and brother.

An unfettered couple again, Diane and Barnard took cruises, enjoyed friends and the beach cities' nightlife, tried out new recipes and new restaurants. A half-year sabbatical in Germany was a pleasant change, one he vowed to repeat. Music, theater, and the arts occupied many of their weekend evenings.

She noticed alarming symptoms just before their long planned Mediterranean cruise but did not want to forgo the trip and so kept silent. After a marginally profitable evening in the ship's casino, she had him feel the uneven, oblate thickening after they undressed for bed. "It's scary. I'd better see Doctor Mack as soon as we get home," she said. The day a specialist confirmed the diagnosis they had sweaty sex and cried afterwards. They held each other close during that and many subsequent nights. It was hard to discern who was the more frightened. Neither could absorb what they had been told, that hers was an aggressive form of the disease, already well advanced. Peering for hours at jargon in the medical library at UCLA yielded Barnard eventualities with no compensatory understanding and, certainly, no comfort. He read much but gained little, except for the searing pain of her absence long before it came. True to his vocation, he drew parallels between her invasive malignancy and the miasma that was overtaking the financial world. As with economies manipulated and mismanaged by politicians, there would be carefully planned treatments to be borne, often with far less real than imagined benefit.

Diane was game. She accepted with grace the fatigue and the unpleasant changes in appearance, the alteration in lifestyle, the tectonic shift in plans and anticipations. Barnard, like a good sailor, steered a ragged course through a rough sea of waiting rooms, procedures, and tests. At the college there were administrative meetings and extra summer classes to teach. In view of the depressing financial climate, there were financial talks to local groups, occasional early morning TV news interviews. Academically as well as personally, it was natural to imagine that understanding the how and why of something would facilitate control over it, would at least mollify fear. Yet, for Diane and Barnard, as for the world in general, the attempts to understand were cursed with the onus of hope.

Professor Cordner's vague anticipation of an unprecedented worldwide financial tempest was of far less import than the proximate certainty of what loomed over the two of them. They canceled sailing the Greek islands, which had been another planned adventure. Barnard's apperception of those years was dulled by the gray fog of their unremitting progression, by the illusion of perpetual postponement. He did not notice time's pace accelerating or mark the ever receding shore of The Book.

She endured the cold, unfriendly tables and painful needles, the debilitating pharmaceuticals and radiation, the transmutation of hope to gut twisting certainty. She lingered. Then she died.

The all consuming foreboding of her being gone had not steeled him for it. The pain and wretched loneliness, however, were quickly followed by a paradoxical sense of relief – there was no longer something to fear. That the financial world was in turmoil was as much helpful as disturbing. It provided Barnard with other issues upon which to focus. He could absorb the evolving news of the fallout from an unprecedented union of the unscrupulous with the instantaneous and share in the angst, thereby displacing grief. The internecine conflicts in the Mideast, the inexorable rise of China, the rebirth of Russian empire, the concentrations of power that lent outsized significance to marginal factions, and the well camouflaged prospect of a collapse of the dollar gave Barnard much to mull over, to analyze. SMC was having critical financial issues of its own. Barnard reassigned and let go, trimming his departmental budget in a parochial version of what was happening throughout the state and the country.

When not in his office on Saturdays, Barnard took to being a weekend wanderer amongst the shops of Santa Monica, the imaginatively named Venice Boardwalk, and the commercial sections of Raceway behind the beach. Like a tourist, he would drive around Brentwood and the Palisades with the top down on sunny afternoons. Like an insatiable voyeur, he prowled the beach cities of Hermosa and Manhattan, the shadowed, constricted alleys of Venice. Imagining he was extracting informal economic data, he sampled people's lives through their stoops, entryways, and patient piles of discards. In truth, his outings had no true purpose other than to distract him. When the weather was poor, he roamed the shops of West L.A. and Hollywood, the malls of the Valley. He mounted repetitive expeditions to purveyors of cheap goods fabricated by underpaid children abroad and mused over the economic implications, most soon to be realized. When the weather was pleasant, he made use of the curvaceous strands of concrete at the near edge of the beach. He strode amongst poly-clad peers and those more youthful who zoomed by on bikes or skates. To escape these intrusions he would trudge across the sand and visit the swash. Often he stood, in long intervals of pensive recollection, atop the cliff at the western terminus of Wilshire Boulevard to feel the coastal breeze and look out over Santa Monica Beach, the statue of the eponymic saint staring inland behind him.

"How about this, Barnard," his father once had said, looking at the same holy figure. "You and I'll get in that little car, that Porsche right there, and drive all the way from here to Chicago on the same highway, number 66. Just the two of us. Wouldn't that be fun?"

"Where's Chicago?" Barnard had asked.

He eventually learned of the error, that Route 66 began miles away, on Lincoln not far from Sunset Vistas. This made no difference. He would have welcomed the adventure, has often dearly wished that it, like other aspects of his father, had been something to recall.

At the college, Barnard dredged up old notes and resumed his regularly postponed project to create a common sense, not abstrusely mathematical economic text. Its chapters on the sociological and behavioral aspects of economics should appear early, he decided, to give them equal ranking with the obligatory classical topics. And it should be aimed at undergraduates, not the lay reader, so as to be more likely to yield stable royalties. Either way, he wanted it to reflect what people actually did, not what 19th and 20th century theory presumed. With the shocks of the first and second post-millennial crashes fresh in his mind, he further modified the thrust of his outline to detail the potential instability of a system dominated by technology that outflanked economic utility.

He stayed at SMC for more than forty years, rotating out of the chairmanship when reaching 67-1/2, as tradition required. The physical rearrangements were easier than the weeding out, the giving up of parts of his past, even a past to which he had only superficial attachment. Shortly after the party, organized by Kyle, to quietly celebrate that Grandpa had made it to 70, the new head of Economic Studies expanded prior hints into explicit suggestion: It was time for Barnard to retire. Barnard had come to the same conclusion. No longer needing to organize lectures, programs, and schedules, no longer having to manage students, staff, and deans, and finally accepting his profound dispensability, he could enjoy the moderate freedom of unencumbered self-absorption.

At home in the late afternoon, he would look at The Machine, recently purchased to keep him in shape and, by displacing at least some of the old, to dull remembrance of Diane. He would stare beyond its bulk through the window steeped in bright sun. After a local meander he would stand in the arched entry of the front room and stare at the couch, noting for the thousandth time the paired chairs bracketing the low table in front of it. After Diane died, he no longer found the couch a comfortable place to read or to watch. Instead, he often sat for a time at the long table in the dining room, both arms flat. Then, reviewing the woodwork of the china cabinet, its glass panels, and the hints of the stemware within, he would idly run his fingers along a shelf and take note of where Alece, his part-time housekeeper, had failed to dust that week but never make mention of it.

One evening, after a solitary dinner at the Krung Thep, still feeling the innocent, dark green morsel that had elicited a paroxysm of hiccups, he turned on the wash lights and walked slowly past the framed prints neatly arranged on the wall opposite the couch. Then, on the other side of that wall, he lay in the middle of the double bed, blanket to his waist, waiting for the faint, quasiperiodic plomps of the surf to be masked by morning traffic. Homes, such as his, less than five or six real blocks from the beach were commanding increasingly high prices. There were no empty lots. The single-family structures, those not already torn down and replaced by hulking apartment buildings, were being bought up by individual speculators – near retirement couples and the successful young – who planned to double their actual cash investment by refreshing the properties then selling to others who would hold for a year or two then sell to other speculators looking to tear down, add-on, or remodel, and then sell to ...

On and on it rolled. The pattern was unmistakable.

The instant the idea arose, selling the house made sense to Barnard. Only, he did not want to leave. He enjoyed where he was. It suited him. And he knew his singular car, by then another small import convertible, also would soon go. He did not want to contemplate giving up those airy rides to nowhere. He wanted no restrictions, no communal obligations. Having a car and a garage to park it in, having Alece vacuum and dust once a week or so, eating what and when he chose, the sum of his virtual if not actual independence – all of that would disappear. When he expressed his inclination to move to Kyle, part of him went cold at his son's prompt positive reaction.

Kyle is an employee, a commodity to be acquired, allocated, and used. His wife still manages the house and the children. Restarting her professional life had been quickly rejected when he lost his job. They instead chose to relocate when, after a half-year of impatient searching, a software management opportunity opened up in San Angelo, Texas – hot, dry, distant, only cinematically experienced Texas. Shortly after their move, they made a wearisome two day drive back to L.A. for a week of visits, first with Kyle's in-laws in Riverside then with Barnard in Ocean Park. It was time for the three of them – Kyle, his bride, and Barnard – to talk. The younger heads nodded and affirmed, shook and denied during that conversation over coffee across the dining table. They wanted to make certain that Barnard had suitable plans and that Kyle would not be needed by his father. Barnard felt an urge to delay the inevitable, to construct an option that, akin to the unconscious motivation of the immature, would shift the burden of choice to others. He wanted an escape clause in case his analysis was flawed and the housing market did not crash. Barnard remembers how it went, or at least, what he heard. He had frequent subsequent occasions to hear it projected from the wall beyond the foot of his bed.

"No. What you said was right. It's time to get rid of the house and all that. We've talked about it a lot.... Yes, Katie, too.... I did contact her, sure.... Doesn't matter. We see the same thing. You need to sell, like you – No. You need to sell the house.... Listen, you need to – Yes, yes. I, we know, Dad. That was only until you made up your mind. Ever since Mom – But they've got several nice apartments available. It'll be like your own place, so don't – No. Mom's been – That's silly. It will be your own. You won't be sharing and you'll – Listen. You'll – Listen, Dad. You'll be able to come and go and – Dad. Listen to me. You can come and go as you want! See? Eat out as often as you want. What?... No, it's not.... Yes.... Yes.... Good. Okay then. We'll, I mean I'll set it up with – Yes. With Miss Chen at Vistas.... Chen. It's cee, aich, ee, en.... Oh, I don't know, Dad. Chinese, maybe.... No, I don't know her first name.... Okay?... Good. We'll arrange a time."

"Make the best of it, Barnard," she had added from the side, away from the focus and lukewarm as they talked across the table that afternoon. Barnard knew what that meant. He shared his daughter-in-law's reluctance to have him be a utile but ungainly appendage to their lives. Nor did he want to exchange coastal SoCal's perpetual spring for summers that are long, hot, and sterile, for cold winters that were distressingly unlike New England's or Arrowhead's. No sea? No beach? Endless views of barren scrub and flat horizons? That and the prospect of being so far from the shore repelled Barnard.

Each day after their tour of Sunset Vistas brought further validation that it was time to move, time to extract and put to good use his substantial residential equity. Professor Cordner's Last Lemma: In good times or bad, money in hand is to be preferred over assets on paper. Nonetheless, he vacillated. Southern California real estate has always been a succession of booms and busts. True, the next bust could be even worse, of monumental scope and severity. Could be, however, was not would be. His inclination was soon confronted with circumstance. A choice, one bedroom unit at the once familiar, now newly enhanced Sunset Vistas on Pine Street was immediately available. It would be fully paid for and secure for life. A short bus or van ride would suffice to get to where he was accustomed to going. The inevitable was abetted by meritorious argument:

Friends? New.

Meals? Provided.

View? High and nice, he had been told. "Of what?" he had counter proposed, receiving a small smile in reply.

Cost? Not an issue.

What if? Assistants would be near, even at night.

The unpleasant issues of aging, so glibly dismissed before the fact, would be taken care of through the annuity arrangement at Sunset Vistas. Even if Kyle had not concurred, the financial facts were impossible to dismiss. Promised Care was gloomily rational. Barnard would have preferred to stay in the home he and his Diane had made their own through years of love, birth, progress, and pain. He did not relish being a tenant. Neither did he wish to lose control over his surroundings. Despite the detailed charts and well-researched parallels that he had presented in class, despite all that he scribbled down for The Book and the reality thereby implied, he loved that compact house. It was, after all, the locus of many fondly recalled, happy times. The immediate reality, however, was that his investments were withering. Kyle had married, started a career, seen it diminished by de facto as well as actual importation of offshore technical labor, suffered through a decade-long economic blight along with the rest of the modestly skilled, then moved out of state. Katie, having hooked up with some skinhead fool, had slipped from his life long before. He was alone.

There are performing assets and nonperforming, just as there are recoverable downturns and there are death spirals. The trick is to distinguish early one from the other or, as is usually the case, to make a lucky guess. Barnard's students would initially find it difficult to grasp the relevant application of expected value. They would take it as a singular, theoretic assignment out of a universe of potentialities. One could always personalize it and thereby be the exception, they imagined. Once Barnard realized he was thinking along these lines, he isolated and reexamined the uneconomic factors that were confounding his choice. He accepted what his emotions wished to reject, namely, that his circumstance unquestionably compelled getting out while he could.

The compact, memory laden house sold quickly. It was before the peak that preceded the big crash, so he did leave a fraction on the table, did not optimize in the short-term. But it was close enough. Barnard accepted moving on to a new phase. He had to. His assets were fixed. He needed the remainder of his life to be as secure and uncomplicated as he could make it. Predictably and consistent with his nature, having made the choice and acted upon it, he came to question it. Confounding the best of intentions, unanticipated insecurities and complications do often accompany an unfolding reality.

Despite the second thoughts, however, he has adapted and grown accustomed to the functional retirement residence on Pine Street. He would have preferred to be located where he had an easier walk to the beach. Moving to a condo on Speedway would have provided that plus easy immersion in the familiar pleasures of shops, surf, tasty bites, and babble. An investment account with its decimal point one step further to the right might have made that possible. And that The Book could have done. It could have enabled him to glean tangible value from his four decades as an academician. He rues having allowed a cascade of distractions to preclude completion and publication before its anticipations achieved the status of fact, before they overripened to be soft fodder for journalists. Above all, he rues his characteristic lack of persistence and resents that he had been burdened with this trait.

Sunset Vista's locale is familiar to Barnard. The area is neither expensive nor exclusive, neither all young nor all old. There are places to visit nearby, the OP, and storefronts to stroll by on Lincoln Boulevard. The Venice Boardwalk is a few minutes away by bus, as is the Santa Monica Promenade. It helps that Barnard is relatively lucky with his health. With good posture and the energy to skip over a pair of steps when he chooses, Barnard, at 79, has changed little since his arrival. This is his perception, at least. He is proud of his appearance. Less so, perhaps, when he sees it in unanticipated reflections. Nonetheless, he cannot suppress a warm feeling in response to a cheerful "I don't believe that," or "My, really?" when confirming his birth date at the pharmacy, checking in with a nurse, or when revealing his actual age to an aide or new arrival. Such banalities, even if often less than genuine, do count for something.

Barnard misses his convertible, misses feeling the wind on overly fast drives along the coast or through the tricky curves in the canyons. With little besides conversation, wandering the local area, and the Internet to occupy his time, Barnard finds food and its contemplation to be welcome, incidental diversions. Max Lohren's Ocean Park Café, the OP, is a prime destination. Max's pressed bread sandwiches, pressbrot – similar to a panino but with a distinctive, interesting shape and filling – are a special treat.

As for transport, the local buses are quiet and, for those over 70, free. Not long ago, when 62-1/2 was the threshold, they still used diesel. The buses roared and stank. Now they hiss and hum, use LNG. There have been intimations – tax-conscious whines of appeasement in the spews and in the Council – that free bus service for retirees must be the next thing to go or that they should increase the cutoff age to 75. Either would be another bureaucratic condescension that signifies much and accomplishes little.

("Fictional Finance, Pretense as Policy" was the title Barnard had applied when his lectures called for touching on similarly misguided organizational, especially governmental budgetary endeavors.)

It is hard to determine what to take seriously, since most of the posted, generally darkly themed, anticipatory rants originate from the insecure or the devious, from manipulative, insular souls whose abundance of spare time far exceeds their wits. The bloggers and the snipers ejaculate impotent text as if they have an important observation to present or, failing even that, some novel interpretation of dubious fact. They seldom have either. Nonetheless, it is true that some change in bus service would be unsurprising. It is obviously slipping. Adverts obscure the windows; half-heartedly erased graffiti offer dubious extensions and sophomoric puns. Maintenance, reliability, and frequency, even driver attitude, have succumbed to diminished funding, in line with the entire economy having changed since that last big crash. The angry conjectures are understandable.

For Barnard there are elements of irony, of classical tragedy in these last twenty or so years. He, the observant academic, had repetitively lectured on academic principles vis-à-vis the actual state of the world. At seminars and invited lectures he had harped, in researched and well illustrated detail, on who would pay for the economic mismanagement and financial misdeeds: excess sovereign debt and Ponzi bonds; bipolar demographics with the middle left to languish as in Venezuela and Argentina once and then again; an emergent China; instantaneous algorithmic trading; specious precious monetary metal and basic commodity pricing; the dark and devious aspects of cryptocurrency; and, of course, the new self-assured China. He was faithful to his premise of behavior not theory being the cause of financial disarray and had many anticipatory warnings to offer to less than raptly attentive student audiences who, for the most part, were eager to get out and get in on it.

Yet, despite what he saw and had divined, despite that, like a competent helmsman, he did what logic and his intended destination had demanded, despite that he had made maneuvers that seemed unquestionably correct, he was blown off course by national insolvency much like a swabbie might be tumbled on a slippery deck. Just six years ago, before Barnard bought his compact set of rooms, he had been disingenuous in putting contradictory financial issues before Kyle. In truth, he knew a serious break in the dollar was coming. One would be orchestrated, in his opinion, to facilitate a tactical retreat from unsupportable fiscal errors. He was certain the attendant turmoil of the first third of the twenty-first century would be unprecedented. Ineffectually prescient, Barnard is paying an errant fool's price along with everyone else. He now must contend with a turbulent sea and an uncertain port.

He knows precisely the tale that needs to be told and would welcome the opportunity to write the story of recent financial history:
- Who was in control.
- What the real problems were.
- Why some imagined they could be managed.
- When fictive, politically inspired regulations, which a foreign third party – even one in deep shadow – can obviate at its whim, cease to be adequate.
- How posturing, media management, and hubris can be effective but only in the short run.

Domestic misadventure has gone on for too long. The situation has passed the critical stage and is beyond the facile, indeed devious management that had been the rule. The necessary realignment will run its course with the glacial speed of most societal change. As recent events have shown, the veneer of the ongoing "re-formation" makes it seem truly fresh, while its core is actually quite old. It is as common and familiar and odious as evil itself. Few of Barnard's age had imagined that their economic prospects could decline so far and so fast, that most of the plain vanilla retirement condos would give up. In truth, he had not either. Much of his retired generation's security, along with the funds they had put aside for it, has been lost nevertheless. The government has stepped in under duress but with duplicitous argument. In the past even this might have served the general interest. This time it has not, because there are more than economic forces in play. There is an undeniable storm of malignant change. Fortunately for Barnard and his small circle, Sunset Vistas is still operating, if barely. New residents do arrive and bring in fresh annuity money to fund current expenses.

Yes, there is a name for that.

The future of those expectant students who received his final lectures, those who sat in offset rows in sterile academic halls, in varying states of attention to their screens and no doubt on their way to minor careers, will depend less on skill and effort than on circumstance. The economic system within which (for which?) they toil, imposes constraints unfelt by their parents or grandparents. That prior trajectory was an anomaly, a few decades or, if generously counted, a half century that defied historical patterns. We are in the initial stage of the Great Revision, some say. Barnard prefers a different phrase: The Great Leveling. Being average is an unpleasant state for those who assumed being above it was their right. If those who have adjusted thus far are not afraid, it is because they are not sufficiently well-read to anticipate the degree of loss that may yet be in store for them.

Barnard is a solid and practical person. He enjoys carefully crafted films. He talks and argues with his two friends to stay alert, to feel involved. He prepares main dishes from memory and cuddles quietly with Sallie. He wishes he had an aptitude for Go or chess but at the same time recognizes that he would then either have to find a like-minded real person or acquiesce to staring at an impersonal display screen. He used to regularly watch "London: The World Today" and read the foreign financial press so as to refine his views. Instead of egocentric, event-driven one-line plaints, he would often composed expansive, explanatory commentaries, which he submitted but rarely saw appear. He had signed them "Barnard Francis Cordner, Professor and Chair of Economics, Santa Monica College." He might have seen his missives better received if he had left off the irrelevant appellation.

All of this is largely behind him. Barnard's world is far simpler. His apartment is compact without being confining. On a good day, being at Sunset Vistas approximates having his own house but without the attendant chores. He has Sallie and a pair of good friends, yet is burdened neither with responsibilities nor commitments. He is on the fourth floor with a partial view and a place to exercise only one floor below. He can wander about as he chooses and has the OP Café for tasty pressbrot. When the time comes, as it must in the best case, he will progress up to Five or Six, be a sedentary, stay-at-home client. But not yet. No. Not for a while. Hopefully not for a long while.

And Seven?
Never.
No Zombie Zone Zeven.
Not ever.

CHAPTER 5

SUNDAE NITE

Sundae Nite draws him in. Barnard takes advantage of and appreciates the small pleasures it provides. As always, there is an additional subtle benefit. The event tempers the fear of the inevitable end that all seek to delay. Witnessing the drama of gradual decay in others makes recognition of its alternative – a quick demise, which is the way he would have it – palliative and, paradoxically, even protective. Barnard would not presume to utter Hamlet's existential concerns, although he feels them. He steps back to the table, grasps the thin-walled plastic bowl firmly, to steady it, and ladles in dollops of as much creamy white and rich brown as it will safely hold.

Soft. Syrup over.
Right.
And nuts. Colon pockets? So no nuts? No. Must nuts. Must.
Such little bowls.

He crisscrosses the mass with several passes of dark syrup and adds one, then a second helping of the chopped peanuts. Belatedly, he retrieves his customary pair of napkins. One will be for his lap, the other to wipe lips and chin. He resumes his search for a spoon, the spoons. Growing impatient, he knows they must be nearby.

"Pfffffffhhhhh."

They ... got ... them ... Where? Ahh.

He follows the lead of his eyes toward the pink plastic utensils he at last has located, nestled on a cloth at the back of the far end of the table. He shakes his head slowly.

Why cloth? They're disposables.

Barnard sets down his construction to await his return from the necessitated short trek. Arriving at his goal, the two napkins clutched firmly in one hand, he leans across to pick out a spoon.

"Should be together," he mutters, his cheeks warm as if from shooting a high tropical sun. "Silly. Plates and bowls there, forks there. Spoons here? Stupid."

His sotto voce complaint, he fears, is certain to elicit another nocturnal monologue or the composition of another never-sent Memo to Staff. He again approaches the figure who, now lounging near the middle of the table, is forking a slice of cake. Disjointed images blossom and overlap as in a kaleidoscope as he steps around. The present dances with the past: snatches of Sunset Vistas of this day; of how it was, of how he got there and why; pieces of ancient events, of what may yet come to be. It takes barely an instant for him to skitter through decades.

I'm here. It's now, damn it. Now.
Not declining. Won't. Like a young sixty-five.

Barnard walks back slowly, deliberately. He blinks to clear away the chaos of competing allusions.

More reps. Tomorrow. Need arms strong, quick. Explosive.
No reason, except ...
Wish slept better. Too much sitting.

He takes a closer look at the casual form in passing. Experiencing a stab of annoyance because of this second, obligatory detour, he sucks his tongue as he transfers the adjudged to have been administratively malplaced small spoon to join the napkins in his left hand. Barnard pauses, feels a slow, deep breath expand his chest. Looking aside at the teasingly familiar figure, he stands immobile. It takes several seconds for his eyes to fully focus, for him to recognize that it is Doug's nephew. He has the vague inclination to greet him, but hesitates.

Right. Two peas in a ... Never really met.

"Hello. Having a visit?" Barnard asks, louder than he intended. The young man studies the utensil in his hand then jerks his head to the side, his mouth busy with a final morsel.

"Ah huh. My Uncle, Doug Roach," he says, after a swallow and quick flick of his tongue for the few crumbs adorning the corner of his mouth. He leans to one side, looking to get rid of his cleared plate. Swiping across his mouth with the side of his forefinger, he moves away, taking pains to mime he is searching for someone or something.

Barnard allows a faint smile of empathetic scission as he retrieves his own carefully created treat. Standing close by the table, he lets his eyes wander as he spoons it up, shifting sideways to look beyond the diminishing nephew to Sallie and May at a table near the wall. He decides to finish his sundae before joining them, thus easing his opportunity for another. He guides the spoon to retrieve the final reluctant, syrup-imbued nut bits. The room seems to recede. Time seems to be passing slowly. He senses that he is glad for having Sallie and for his Sundae Nites.

Hope never cancelled. Never.
Tired. Early to bed tonight.

The future is never set in stone. It can be preconfigured or dismissed entirely, as one chooses. No so the past. The past is a heavy, fatiguing, controlling intrusion that must be willfully overcome, lest it overtakes, like the press of dark water in survival drills. Barnard had felt it, earlier, as he approached the open double doors of the activities room. He had detected it in the scent of the air moving past him into the dim hallway as he stood at the entry, heard it in the poorly parsed snatches of conversations as he engaged this Sundae Nite, saw it in brief sparkles of refracted light.

While Barnard has progressed beyond resignation, he has not fully attained acceptance. His sense of ease, which arises from being where he should, is diminished by the nag of not being where he would prefer. The Shaker Hymn states how it should be. That is the plan. Perhaps even Relocation, moving to a remote Elder Eden, were that to happen, will be tolerable. The alternative is to be dead, or to be unreachable as may be those – unnameable yet tenuously familiar memories – whom he no longer sees on his Venice walks or at Lohren's OP Café, or who have gone missing from Sunset Vistas. Barnard fights time, will not allow the reality of these final years, which is the objective truth of it, to take total control. However, finality is inescapable, its schedule increasingly acute.

He flexes his shoulders, makes that familiar effort to be buoyed by the present. It is another second Friday evening, another Sundae Nite at Sunset Vistas, with replays of old conversations, coffee or tea, sweets, and – what is primary for Barnard – ice cream. He is here, extant and active even as he closes in on his four score. That alone should be a significant achievement. There should be some reward.

Sallie is precisely that, he is inclined to think. They enjoy patient closeness in the evening. They share their reactions to what often masquerades as news but that, in light of the unwanted misdirections, Barnard has christened "spews." Frequently they share his bed, enjoying being close and warming each other. Unfortunately, she does not care for beach walks as does he. And they have stopped going out for dinner, in part since it is less open to them at night. Finances, also, have played a role in their decision to be less adventuresome. Still, in many ways, it is as if he has been provided with at least a symbolic vestige of a pleasant past. Without doubt, Barnard is very glad for Sallie being part of his life.

Only, she does not need to keep watch over his appearance, he often has grumbled. Earlier was a good example. His haircut is just fine, he asserts inwardly as he stands watching the Sundae Nite mélange. He needs no approval. Like his impatience and his marginal auditory deficit, her occasionally excessive concern is really a small matter. It does not intrude upon the pleasure of her being here to share Sunset Vistas with him. He would love to jump into a car with her, put the top down, and take a fast drive, up through Topanga Canyon perhaps, or even just a short ride for ice cream. Unfortunately, even apart from his having given up the convertible and their need to limit their sugar intake, the pleasure of a crisp cone or dainty glass dish is difficult to arrange. There is not one true ice cream shop easily accessible from Sunset Vistas. The last holdout, Lincoln Creamery, once only a short walk beyond the OP, had finally given up rather than adhere to the strict California Obesity Notification requirements. Neither chemically flavored yogurt nor the "healthy" ersatz ice cream, in hard, square cartons at the market, yields the same pleasure. The Graese Dairy's offerings, exploiting the loophole afforded private institutional purchasers, are extravagant treats not yet changed by fiat. Barnard therefore enjoys, especially looks forward to Sundae Nites.

His favored weakness goes back to a lifetime ago, when speeding along in a chattering 356 Porsche, with a parent too soon taken, to find the "perfect double." Thomas, Father, was enthralled by that unadorned exemplar of German engineering, one of the first of its kind in Santa Monica. He could not have enjoyed it more even if it had been bought with his own money. Adolph struggled to restrain his begrudging admiration of the vehicle and its hints of validation. He would not have tolerated a Jaguar or MG roadster; an Alfa, possibly, or a Ford convertible, even after the war, since its namesake's *The International Jew* provided many good quotes for his speeches.

With cloth top lowered and secured, father and son would head for the Hollywood Freeway, swoop into the Valley in the left most lane. Muffled by neither water jacket nor heavy metal, the air-cooled engine made a unique metallic buzz amongst the throbbing ease of glitzy Chryslers, black Buicks, and Beverly Hills Cadillacs. They would exit the freeway to speed up a narrow canyon road to Mulholland Drive, playfully slaloming left and right along the latter's winding, relatively undeveloped course. They would cut south above Westwood, then take Sunset toward the Pacific. On route up the coast highway, they would turn off into another twisting, narrow road for their ice cream stop, which would allow the tingling in Barnard's feet to subside and give him a chance to push the hair out of his eyes.

Otherwise motionless, Barnard lightly passes fingertips along his graying temple. He can almost feel that ancient vibration and the flutter from the wind, a sensation comparable to the shaky hypoglycemic buzz that will now often accompany too much sugary ice cream. It has been several hours since the solid food at the funeral. It might be best, he decides, not to have that second sundae this late at night.

Father.

* * * * *

Thomas could eat huge portions of ice cream because he played tennis several times a week and easily burned off excess calories. Barnard's mother, in common with the other substantial European women of the family, found this trait to be unfair, as did Adolph Graese. His grandfather was troubled more concretely by Thomas's frequent midday absences for matches and his Friday expeditions with Barnard. Regardless, their favorite ice cream counter – in a shadowy, mysteriously fragrant general store as Barnard recalls – was visited regularly. There Father would encourage him away from plain vanilla. He would deem Barnard's simple choice insufficiently daring, even if with a slip of chocolate. "Bore Ing," he would state with a grin. He would cajole his then singular son to adorn it with something, say a scoop of peach and a mass of sticky walnuts, or to perhaps try adding rocky road and a drizzle of strawberry sauce. The suggestion of a thread of marshmallow cream and/or a dot of bright red cherry finally brought Barnard to the edge of adventure. Rich chocolate in a paper cup, under contrasting French vanilla covered in syrup and adorned with a heaping of nut pieces, was his eventual firm preference, which was repeated on many subsequent visits. Located along a narrow road seemingly not far north of Santa Monica, the potentially memory-laden local market eluded Barnard's later attempts to find.

Barnard has inherited few paternal traits that he can specify except for that love of ice cream. Occasionally, as when talking to his grandkids, there will surface snatches of the stories Father read to him. Some seemed to have had particular significance, an extra essence that prompted patient, often textually incongruent commentary that Barnard can only vaguely recall. Of his father as a person, Barnard has even less. One vivid image is of his parent's anger at having his son greeted with the diminutive "Bernie."

"It's BAR Nard, not BER Nard. There's a difference!" he can recall Father putting sharply to the short, heavyset man who stood deferentially aside his desk.

Barnard had followed behind as Father retrieved forgotten paperwork from the main office before their outing. The hawk-nosed clerk meekly restated his greeting.

"Hello, Bar Nard. Yes."

The reprimand was overly strong for such an inconsequential matter. Yet Barnard, with subtlety modulated insistence, perpetuated that same distinction once he entered school and began to meet new friends. It was about then that the patriarch announced his decision to indulge his devotion to fringe politics. Finally, A.G. decided, he had worked long enough. Thomas would "step into the Old Man's shoes," managerially at least. Barnard's grandfather was not inclined to hide his disappointment that the land of his birth had not prevailed in the last big war. Many in inland California, where the dairy's actual operations were located, were sympathetic to his far right arguments, to their intentions if not their facts. Adolph, who indeed had big shoes to fill, had the means as well as the impatient conviction to transmit his fantasies of political change. And so he did, until the fact of his own mortality silenced him.

Adolph Graese was rigid and spare of warmth. The success of the Graese Dairy was due less to his enlightened management than it was to the underpaid, pliant, half-educated migrant workers upon which it depended and whom he detested. Being in charge, A.G. could claim that the precise opposite was the case. As he rejected society's burgeoning liberalism so had he not welcomed his elder daughter's choice of a marriage partner. It was another judgment he was disinclined to hide. Only upon her insistence had he taken Thomas into the family enterprise. As the new General Manager, Thomas did about as well as had been anticipated, which was not very. A.G. made his dissatisfactions explicit on many occasions, as governmental regulations and competition (the significant realities) and the lack of a firm managerial hand (his self-serving view) exacted their toll on the dairy. A misguided foray into the raw, unpasteurized dairy product business was particularly disruptive. Barnard can faintly recall flashes of the agitated discussions – Thomas and Mother on one side, Aunt Laura and Adolph on the other – during discordant dinners. While he was too young to understand the details, he had no difficulty grasping the acrimony. His Jane-Jane would observe without taking sides. Barnard regrets that he knows little of those years and nothing of what his father was trying to achieve.

One pleasant Friday morning, before their scheduled outing, Barnard's father drove into Westwood. So that "his doctor could take a look at his shoulder," his mother later explained to Barnard. "It was probably a bursitis," she said, a condition that should be treated before it got so bad that he would have to stop playing competitive tennis. Thomas disliked taking any medications, even aspirin. He would be more likely to agree to using that, or perhaps a stronger anti-inflammatory, if officially told it would be helpful. While there, and with no additional forewarning, he felt his chest clamped in a vise of pain so profound he could not call out. There was the frantic dash of the crash cart, soon followed by an agitated telephone call to the house. Father was gone. Frances was not yet certain she was pregnant.

Barnard and his father had been allowed insufficient time. His appreciation of the significance of his loss came when the ice cream outings ceased. It was punctuated by the absence of the noisy Porsche. He was nurtured thereafter by Mother, his Grandmother, and, later, superficially by Aunt Laura, who joined them in the big house when Adolph died, not many years after Thomas. Father was, therefore, an abstract construct to Barnard. Any hints of emulation were instinctive and unverifiable, never noted and never reinforced. His brother, Anders Kurt Cordner, seven years his junior, provided no linkage. In fact, having made his appearance after the death of his father, Barnard felt intruded upon. He resented the affection that flowed from Grandfather Adolph to the tenuously miraculous, properly uncircumcised junior Cordner, one who was deemed to have a greater likelihood of being molded in the old man's image. This was to be Laura Grease's self-assigned but unrewarded task.

5 - - SUNDAE NITE

Neither Barnard nor his mother felt sorrow when the old man passed. They were, however, financially disadvantaged, as were the many fringe groups that he had supported. The latter missed Adolph Graese bitterly, as did Laura Graese and Grandma Jane-Jane. Insurance, supplemented by their minor fractional interest in the dairy, yielded little to Barnard, his brother Anders, and his mother beyond that of her not needing to work or remarry. With Aunt Laura in control, the five of them could make do in the big house. Both brothers would receive fair starts through modest trusts that had been established for them. Anders – impetuous, strong-willed, and volatile – had little patience for higher education. He absorbed some then quickly focused his attention on the stock market and speculation.

By Frances' late sixties, with Anders dead at his own hand after the stunning but transitory market collapse of '87 ("Gone. It's all fucking gone," were his last words to his brother) and Barnard off on his own with two children, she had had enough of the repetitious visitation of familial recriminations. She moved into the Eastern Star retirement home, where she was an honored guest. Their mother, Jane Graese, passed not long after, leaving his Aunt Laura to care for herself, which was not much of a change.

At Eastern Star Frances Graese Cordner resided amongst "old people," aging souls who watched the various screens and each other, who waited for visitors, meals, or new arrivals, and lamented, defamed, or simply recalled the departed, all depending upon mood. They talked, not infrequently to themselves. They had crafts, always in groups and usually choreographed so that a wheelchair or walker was no hindrance. Visitors spoke on travel, health, or world events. There were holiday get-togethers, birthday parties, tenuously themed special events, and religious observances. There was much religion since, for many, this was all that eased the dark certainties ahead.

Frances remained apart from it all. She only tolerated those last years and never seemed to fit in.

It was again a notable parade that final weekend at Eastern Star, a busy flux of visiting relatives. Barnard was sixty-three. It would be the last time he would visit his mother. Consistent with recent practice, he came alone. Diane, never on the best of terms with her, by then was seriously ill and could less easily feign warmth. Mother understood and occasionally inquired as to his wife's health. It was easier for her to accept the *raison d'jour* than to try to extract additional comment; Diane would outlive her by only a few years, as it turned out. The circulation of in and out was a familiar part of each of his dutiful visits. Upon arrival, Barnard would drift in on the mid-evening's tidal flow, navigating slowly amongst the clusters to extract sequent, truncated samples of their conversations, one side invariably difficult to discern. Like an experienced submariner manning a sonar, he could extract weak signals from the noise.

"And how are *you*. Doing okay? Did you have a good dinner?"

"... improved, yes. What?... No, he didn't, no. He ..."

"We'll stop by next week, Daddy. Phyllis has a ... "

"Yes, all right. Both of them, then.... Love you, too, Daddy."

"... was last month. Didn't I tell you? I'm sure I did!"

"Give Nana Betty a kiss, Carrie. Carrie! Give Nana a big ..."

For those arriving, it was waves and warm smiles once their target was spotted. Old eyes, in expectant faces, followed them in. For some, evidence of recognition would appear only after they were approached. There were those who, even though relatively alert, were already dim to outside memory. Few came for them or were expected. They visited among themselves or discreetly looked on sideways, as if spectators to a confusing pageant. And there were those not even so marginally fortunate. They were the blank-faced ones whose gazes were never receptive. But this was infrequently witnessed. Residents of Eastern Star's CCU, the Cognitive Care Unit, then limited to the fourth floor, were not brought down for their visits. These would take place in the security of their own rooms, where they were immersed in unilateral heartache, where no one need suffer shame or embarrassment.

Riding on an opposing flow would be the smiles and waves decorating the ebb tide of those in neither hurried nor reluctant withdrawal. Old eyes would follow them out. For the young there would be those few seconds in the elevator, then their dash out before they were in their cars, anxious to get on with busy evenings. Even for the adult visitors, there were conflicting intentions. Next time they would stay longer, they vowed. That was always the plan. Barnard had done the same, his promise ungenerously burdensome, its improbability familiar.

"She looked pretty good today, didn't she?" one of the departing might offer en route.

The media units would be switched on, the earbuds inserted.

Search. Select. Stop. Search. Select.

"Why do we all have t'come? They make such a fuss over cake n'ice cream."

"Why does it have to be in a big room like that, up on the third floor? All those old people. Should be downstairs."

There were reasons, reasons they would not understand until their time. The visits were rituals of connection. Their general course, witnessed by many, in addition to the specifics, shared by a few, were equally anticipated. Their value rested as much upon communality as upon familiality. The running tides of lone souls, relatives, and descendants would play out their respective roles in a multi-generational drama of both peripherally shared and directly experienced renewal.

Frances Cordner, Barnard's mother, was at her usual spot, well away from any drafts, when he reached her for that last visit. At eighty-six, she no longer was obliged to acknowledge the obsequious solicitations to which her maiden name entitled her. The wheels inserted into the front legs of her heavy armchair allowed Barnard to easily adjust her orientation, as she always requested. Distant financial misadventures had ripped the heart out of a vibrant economy, had enriched a few and savaged many. It was a temporary setback, or so it was opined at great length. While the flesh and blood, the durability of Arab solidarity east of the Mediterranean were being tested, the Muslim Wars had not begun in earnest. As usual, Barnard referred to these ongoing events and offered his opinions. His mother, while lucid to the end, correspondingly showed her typically faint interest in current news. Instead, their conversation revolved around matters essentially inconsequential and often ancient.

Shifting his attention to either side as they talked, the recorded music, and the almost oppressive heat that the elderly seem to require, barely registered on that particular late spring evening. School was out. A larger representation of older offspring – teenagers – was in attendance. Their vigor and independent deportment made them stand out among the usual influx of middle-aged visitors with children, grandkids, and often great-grand babies, in tow. They were beyond innocence but had not reached the age at which communing with an aging relative offered that spice of love or guilt to make it agreeable. Barnard took note also of the spontaneous segregation of the older boys and girls. They were not slyly chatting each other up – boy/girl, girl/boy – the way Barnard might have expected of high schoolers. Surely they were not strangers, he thought. Rules must have been set down before they embarked on the visit: It was not their party; they were not there to socialize amongst themselves. Negotiations had no doubt preceded their arrival, with appended promises of prompt release and infrequent repetition. Still, those pre-adults would remain comfortable around their respective targeted residents for only so long as easy salutations were being passed. They possessed too few significant shared moments, too few lines of true contact. They evinced no awareness of a sequence that foreshadowed their own lives.

The smaller children were a different species entirely. Equally uncomfortable, they could not mask the slowness of time. The single-digit boys were especially active. Their short legs ran even while sitting. "Why aren't all grandchildren little girls?" he considered as they talked, inconsonantly thinking of his own grown children and the nature of fate.

He received the late night call midweek, while proofing the syllabus for his summer short course: "State Deficit Abatement: Bureaucratic Risk versus Reward." His mother died still remembering. His visits, often postponed, had been insufficient, fractional hours of filial contact. This troubled him for many years.

* * * * *

Barnard experienced the conflicted emotions then, had lived with them. He feels them again as he stands silent, solitary, and momentarily sated. Mother had spent almost a score of years at the provident Eastern Star residence. She, too, had sat quietly, expectantly, except not on a vinyl covered chair near easily maintained, cold composite tables of the sort he sees scattered about this evening. Her circumstances were more tasteful. Still, her residency was too long a span of routine and solicitude. Barnard would prefer to not be similarly warehoused, a passive something to be viewed or visited, albeit with a modicum of respect. While he has formed no coherent plan, he has determined that to Go and Be Done With It would be best. He has firmly rejected that fictive graceful slide, that noble, poetic "giving in to sleep." For him it will be a running leap off a steel beach or an "All stop!" headlong dive from the stern. He will grant no desperate grasping, no futile clinging. He had explored those waters, had sailed upon those currents. Having been witness to it once is sufficient.

A few years before Diane.
What if she?
She didn't. Right.

Barnard walks away from the serving table, blends into the murmuring assemblage. The individuals within are barely animate, seem to move in slow motion. Barnard's recollections blend with the present, so that Eastern Star as it was reverts to Sunset Vistas as it is. This night he feels another pang of regret that biographical legacy sessions had not been available in time to capture a detailed remembrance of those close to him. He should have paid closer attention when his mother was alive, and even Aunt Laura. All that he has are a few hours of Frances randomly reminiscing in an old video. There is nothing at all of Father or the rest. For him, certainly, his father, there had been no pressing need. After they were informed of Diane's diagnosis, on the other hand, need was evident. The means were available, but creating a documentary Biograph would have been too much an acknowledgment. They instead occupied themselves with sweet reminiscence while watching boats at Del Rey and listening to the timeless sea, with quiet dinners out, with speaking of vague, painfully impossible plans.

Barnard completed his own extensive series of Biograph interviews shortly after selling the house. He had made the effort, somewhat reluctantly but without overt encouragement, telling himself he was doing it "for the grandkids." That immediate descendants' curiosities are often blunted by more self-centered concerns was exceedingly evident. He could not refute the corollary that most sense the importance of the lives of those bound to them by genes and/or circumstance only when their own existence is about to run its course, when comes the realization that the accruals of time do end. Therefore, in several impromptu sessions he had reminisced about his experiences, about his accomplishments and plans. The result would be more faithful to his self-selected reality than would be secondhand memories evoked by media stashed in unfrequented drawers or packed away in sketchily labeled boxes. He inserted recollections into the interviews as they came to him, trying to appear spontaneous. Images from old photos were added later, some from his archive, some from Kyle's.

With VieGie's automatic updates and data collection, compiling a Biograph is virtually automatic, a side benefit of its core function of connection. VieGie is up and running to provide thoroughly satisfying contact among family members and friends, even if distant. When a personal visit cannot be easily arranged, VieGie is the perfect substitute, comparable to a video conferencing session but far more comprehensive. Linked to the message systems as well as to the not so subtle oversight of all Internet activity – the googly monitoring to which everyone has grown accustomed – and augmented by virtually infinite memory, it is admirably effective at mining current personal information, even resurrecting retroactive linkages for automatic collation.

The name derives from its French appellation, *Visite la gagiste*, which has been shortened to VieGie, "Vee Gee," as spoken. Young wags put a different spin on it, calling it Virtual Granny, since the technology is used primarily for family interactions and obviates spending a valuable portion of a weekend in transit. Barnard has scant familiarity with the technical details of the system. From the name, he surmises that it initially was rolled out in France. In view of their cultural predilection for reverence of ancestors, he would not be surprised if Asian technical groundwork played a role. Here, in the U.S., it is implemented by the federalized big data servers, ultrahigh-speed networks, and super dense storage utilities. That massive system of national intelligence – domestic informatics – enables VieGie, which in turn makes Relocation a practical arrangement, something acceptable to those who are being invited to forgo the immediacy of friends and family.

Casually adept, Barnard distrusts the über-attentive Internet and, by extension, its adjunct, VieGie. Formed from the bones and flesh of networking, the facile dependence upon teleconferencing, holiday E-cards, and personal messaging, with a bouquet garni of national security surveillance and Peeping Tom added for good measure, he judges that VieGie is an intrusively perceptive, all-seeing golem, a creation with emergent properties that overshadow its stated utility as well as the incredibly numerous underlying transactions. It is just as well, he has come to feel, that it is pervasive only now, when he is old.

Blinking slowly, becoming again attuned to his surroundings, Barnard's thoughts turn to Sallie. With few memories in common, they still find much to share. That she is younger has no significance, so long as they both stay healthy. Without conscious guidance, he has strolled up to her and May, who are leaning into each other as if sharing some secret.

"What are you ladies up to?" he asks.

Pulling apart, they look up, May as a friend, Sallie as more.

"You're wet," Sallie says.

"What? Oh." He lifts his arm to check his sleeve. "Probably from ... uhh ... from the ice ... uh, under the ice cream."

"There?" May adds.

Barnard keeps his elbow flexed as he raises his eyes to hers. His face feels as if it has gone pink again. He looks down to verify that the front of his pants is not actually spotted. It could have been.

Tissues. Good.

Neither is there a dab of his already consumed treat. He mimics May's tight smile with his own.

"Ha, ha," he grins.

It is a tease. The damp mark to which Sallie has referred is on the front of his jacket. It must have happened when washing his hands in the lav.

"It was Douglas's fault," he explains. "At the head."

"Who?" Sallie asks.

"Doug Roach. I was, uh ..." His words trail off. Some things are not worth declamation.

Cheap, workmen grade paper towels.
Shit.

"You're not having cake, are you? It's chocolate. You shouldn't," Sallie inserts to advise him. "And you've probably had enough ice cream for tonight."

Barnard aborts his incipient frown. He looks at his empty dessert bowl, the cleaned spoon within.

"Right, Sal," he says. "No, I'm, uh ... No. That's it for me." He crumples the used napkins upon the sticky residue before dropping it all into the nearby trash receptacle.

"Did you talk to Frank at all?" Sallie asks.

"Today? No. Saw him earlier. Why?"

"He was looking for you. Weren't you talking to him before?"

"I saw him. Right," Barnard replies with a glance back. "He didn't say anything, though. Did he want something in particular?"

"Don't know, Babe," replies Sallie. "He'll find you, I suppose. Are you going to have another sundae? You probably shouldn't. But I don't want you to put on a pretend sad look later and be sorry you didn't."

Barnard does not remark on her obvious inconsistency. The slow movement of his lips outward and his silent nods nevertheless confirm that he has decided he has had enough sweets.

"Ummm," he hums then politely asks, "Would you want me to get you some?" He mimes his query to each in turn, hoping to be given an excuse to cancel his own well-intended restraint.

"We did already," May says for them both. "You saw us. We had trouble finding the spoons, too."

"Pfffh," Barnard emits and shifts his stance.

5 - - SUNDAE NITE

"Why do they do that? I mean, why do they scatter the bowls and plates and everything all over the table, instead of putting them in one spot?" May has pointedly directed this to Sallie.

It is, Barnard quickly realizes, another tease. He gives back only a breathy sigh. May's future does not particularly impact him. Sallie is a major factor in his anticipations. Not so for May. What happens, happens. He looks back at the table laden with hot drinks and sweets.

"If you are going to have coffee, make sure it's decaf," Sallie advises. "No, I'd better not," she adds, in response to Barnard's raised eyebrows and offering hand.

"Right. No coffee tonight," he then affirms, having previously decided.

"Nothing for me, thank you," May says, in respect of the lingering offer.

Barnard looks toward the treat table then to a singular gap in the crowd, a clear space like the lacunae that can take shape in the foam at the edge of the surf or an astronomer might discern in the distribution of galaxies in a broad swath of night sky. It envelops Condi, who is looking faintly into the distance, as those with poor eyesight often do. She seems so alone. He feels good for having visited with her earlier.

Oil and water.

"That's a nice suit." Sallie says warmly. She cocks her head as she looks up at Barnard. "And I'm glad that you got a haircut. You needed one."

"Right. I did," he replies. Recollections of Condi fading, he lightly fingers the button of his jacket. "I might leave early, Sal. To change out of this suit. I should have, before." He locks his eyes on Sallie's. "Are you coming up? Or should I come back down?" He would prefer the former, he endeavors to convey.

"Oh, it's almost time. I'll come up."

"See you then."

Before he can leave, May touches his sleeve. He suspects she has been patiently awaiting an opportunity to resume her conversation with Sallie. Barnard is not distressed. He has neither an inkling of nor interest in their current topic.

Money. Her money and what may be left of it.
And Relocation, no doubt.

"No, don't go, Barnard. Sit with us," she requests. "I want to ask you about Relocation. You think it's a good idea? What's going to happen to the money I put up here?"

Exactly. Two for two.
Children's business, not mine.

Barnard remains standing. He wishes he could offer her, the two of them – all of them in the room, for that matter – a proper answer. He fully grasps what has happened. Knowledge is one thing; how to make good use of it, however, is another. He has little wisdom to impart, other than that her anxiety is well founded. The potential benefits of Relocation to an Elder Eden could be illusory; the promises may be neither reliable nor sincere. In contrast, it is fact, not supposition, that Sunset Vistas is not what it was. Soon there will be little here to rely upon.

"Have you talked it over with your boys?" he finally asks her.

"We've talked," May replies. "What they're worried about, mainly, is the annuity here and my other money. Whether they'll be able to transfer any of that to Chile, when I ... Anyway, that sort of thing. What I'm concerned about is what's going on here. I don't want to leave my apartment. Do you?" she directs to Sallie.

"Certainly not. Barnard and I still enjoy it here. Don't we, Babe?"

"Pfffhhhhh," Barnard sighs. "It's, uh, it's okay. Less than it was, right. That's been unavoidable. But it's okay." He wants to be honest yet sound optimistic at the same time. He looks over his shoulder at the somewhat diminished crowd. "It looks like some have already relocated."

The two ladies mimic his glance and indicate their agreement.

"The ones who kept to themselves, mainly. Like your friend over there," May says. She motions not so discreetly toward Condi. "She's the last, I believe," she says in a low voice. Then, addressing Barnard directly, before he can react, she adds, "I mean, that's fine if a person dies or wants to leave. I don't want it to be that we have to leave here if we don't need to, don't want to."

Need to. Want to. Right.

Barnard shrugs. "It's all the same," he says. "Isn't it, May? Want to; need to? It comes down to the same thing. Right? If it's better somewhere else, why not?"

"There are a lot of them that I'd be perfectly happy to see go, truth be told. I just want to make up my own mind about it. You expect it'll change here that much, Barnard?"

"It has already."

Barnard pauses to see if Sallie wishes to offer any confirmation. She nods a "Yes."

5 - - SUNDAE NITE

"The meals aren't as good," Barnard goes on. "That's obvious. And there aren't as many staff about, not so many things going on. Look, the ones who made a point of showing off that they had plenty of money have already gone. Maybe they transferred their IRA and K bonds to a consolidator, got what they could, and relocated. Maybe they decided that a new Elder Eden has to be better than an aging Sunset Vistas. Right? No surprise there. Besides, there are some lucky enough not to care, who don't or won't notice a difference," he laughs. His attempt at humorous sarcasm falls flat; neither in his audience responds.

* * * * *

The ongoing financial disarray has been seriously disruptive. Those on fixed assets are anxious. The faltering of the economy then the collapse of the dollar and the treasury market have left retirees with less than they had anticipated. They had forgone immediate pleasures in favor of future security, only to end up with neither, or, at best, with both at risk. Had this been the general evolutionary pattern, humankind's frontal lobes would never have expanded so dramatically. That portion of the brain's role of managing deferred satisfaction would not have provided a competitive advantage.

Social Security benefits have been reduced in nominal terms. Furthermore, through inflation and the duplicitous lack of indexing, they provide less in spendable value. This double blow has been masked by the issuance of ultra-long Social Security Bonds, offered at significantly less than par. In practical terms, these promises of future payment are reminiscent of the olde-time company scrip and, in modern times, comparable to what several states, countries even, have again had to issue in lieu of actual disbursement. One result has been the appearance of a loose federation of bottom feeding manipulators who extract unconscionable tribute via acquiring the bonds – at a hefty discount, of course – from those with immediate needs. They are like the front-running skimmers in the once novel high frequency trading crowd. Rapacious behavior of that sort is neither new nor unexpected. It is as common to prominent financial institutions as it is to furtive vulture opportunists and unreachable offshore scammers. Dark pools – unregulated, secondary markets, often using cryptocurrency – provide the tacit means of making specious, even illegal activities safe as well as profitable.

5 - - SUNDAE NITE

The U.S. became addicted to deficit spending. Social theorists touted it and political longevity depended upon it. National debt soared. Beyond bad luck or poor judgment, deceit – administrative and congressional – is the true cause of the marked diminution of what retirees had so carefully put aside. After sequential, laborious, thinly disguised postponements that merely aggravated the situation, it came down to a matter of being unable to spend what was not there. Maintaining the system had been a magician's illusion, if interpreted generously, a Madoff scheme if not. Social Security contributions were used for current operations. Now the secrets lie exposed. Payout delays were unavoidable, as likewise were the caps and the reductions.

Barnard had lectured on this also. "Necessary and Insufficient" was the pun he would call up when presenting this as a hypothetical to students. "'Full faith and credit' will soon be transformed into 'Deferred until you are dead and gone,'" he had said to a group of neatly attired Rotarians. They laughed only in part because of his tight grin. They laughed primarily because they could not conceive that his cynicism could be serious or even argumentative. Barnard's expression, however, was not because it was a false fear, a bogeyman, or naively risible. He had smiled because he knew no one wanted to consider seriously the crumbling foundation. They had attended so as to hear of grand edifices, were hoping for an investment tip or two.

After the collapse of the dollar, because of the difficulties arising from China and our own faltering federal debt service, one of the earliest acts of the new administration was taking control of the pension funds, i.e., absorbing the IRAs and Ks. The underlying justification spanned the entire spectrum of political argument, from feasible through expedient to necessary.

Fact: The quickest way to get money when needful is to take it from those who have it. In the civil sphere, this is theft. In government, it is called inflation or expropriation or, worse, since it would pretend to provide ample justification, exigent circumstances. Calling it fiscal triage also made sense: Preserve the limited resources for those destined to survive and to productively consume, rather than engaging in promissory, egalitarian, redistributive, or other fiscal folly.

* * * * *

His thoughts disorganized, Barnard judges that his long-standing assessment would be too lengthy to convey in this setting. He feels the depressing weight of only a general understanding of what is happening. Although he has observed enough events and gleaned enough facts to grasp the underlying financial issues, he might sleep easier knowing less. All Barnard could possibly offer to May, on this otherwise pleasant Sundae Nite, are a few terse sentences to the effect that this is no anomaly, that there are historical precedents, some fairly recent. Argentina did it, less than a generation ago – in 2008, to be exact – with Venezuela taking a similar path a decade or so later. His perception is that the chaos is not really the result of poor planning, international manipulations, or misfortune. Nor is it solely an unintended consequence of opportunism at the margins. All of these are present in abundance, without doubt, but as pitfalls and tactical tools, not necessarily causes. What if, he is tempted to suggest, the chaos actually has been orchestrated from the start, is part of a patient strategy, the creation of a clever cadre of manipulators who initiate, manage, then ultimately make good use of crises?

How It Will Happen and Where It Will Take Us. Barnard should be focusing his efforts on that rather than the much larger task of an academic text for undergraduates. It is too late for his sketched out introductory portions to be prescient. Events have overrun prediction. European states, led by Italy, and South America, epitomized by Brazil, have already suffered. The remainder of his thoughts, his anticipations, could be fresh, perhaps even compelling by virtue of the possibility of being proven wrong. For those in retirement residences, such as Sunset Vistas, there will be additional restrictions and cutbacks, harsher impositions. The specifics remain hard to discern. He would need not strive for certitude. Even the moralizing Dickens gave Scrooge's ghost of the future wiggle room, had it present that which could be, not what must be. In view of what has happened elsewhere, the retired economist imagines he could summarize the domestic morass this way: Our Government has been compelled to enlarge its role as "funder of last resort" to an unprecedented extent; the resultant financial obligations exceed its capabilities; fundamentally new measures are necessary. Or even terser would be: Urgency compels rationale; rationale impels policy; policy creates means.

He looks at May then at Sallie. Yes, he thinks, it is as simple and impenetrable as that. But he is not ready to start the hour lecture that any cogent explanation would require. Nor will a few short sentences suffice.

His views are far from a final analysis and too complex for a casual evening. Nonetheless, May is expectant.

"Relocation? I don't, uh, I can't give you a decent, quick answer, May," Barnard finally chooses to offer. "Too much background. So many things have gone on, here as well as abroad. There are many parts to the problem. I, uh ... I'm, uh, I'm trying to understand it myself."

Barnard has noticed that Phil is walking his way. With him is Martin Stoole, who is officiously tapping the cover of the binder presumably retrieved from Chen. Barnard welcomes the distraction. He has already typed in the outline for a relevant chapter in The Book and will not add anything to it tonight.

Serially smiling at Sallie then May, Barnard lifts his shoulders to sign: What you see is what there is; explanations won't change it.

"Barnard, hey." It is Phil's typical greeting. Still in full stride as he comes near, he adds, "Let's have lunch Monday. At the OP? Early. Eleven, say. Or quarter after. Okay?"

"Sure, Phil. Right. Sounds good. I'll, uh, I'll meet you there," Barnard agrees.

Barely slowing his march, Phil bobs his head to acknowledge. Martin raises a free hand as noncommittal greeting. Their momentum intact, the two continue on their way out.

Barnard has observed that Phil is enmeshed in the affairs of Sunset Vistas, aspects of which he often shares in sketchy outline. Today he has been intent upon Martin Stoole. Those two plus Chen make a strange triplet, indeed. Her job cannot be easy during the ongoing transition. They could be helping her in some way.

Martin is another story – always superficially serious, always with an agenda. Even tonight, Barnard intuits.

More Current Affairs.

There is much to recommend the remedial, optimistically informative Current Affairs gatherings that Stoole provides. Thanks to single-issue Internet sites and blogs, not many years will have to pass before there will be no one left who remembers what real debate or argumentative – as in potentially verifiable – discourse was like. Once past the current transitional phase, the revival of Skinnerian behaviorism will be complete and such comparisons will no longer matter. Until then, for a stubborn subset of the still sentient, they do. Martin remains on the faculty of SMC and by now should have been assigned significant administrative responsibility. He has not, however, as far as Barnard knows.

Ultraconservative and definitely of the far right, he has devoted too much of his time to nonacademic activities, especially those in support of the recent realignment of political leadership. Barnard intuits that Grandfather Adolph Graese would have found him useful.

Self-important twit.

Barnard had hired a young Dr. Martin Stoole primarily because he felt the department was over endowed with old fossils, Prof. Cordner included. While not as intelligent a person as Barnard had hoped to find, he was didactically benign and acceptably academic even with his unfortunate name and virtually absent chin. A newly graduated Ph.D., Stoole was glad for the opportunity at SMC. It had not worked out totally as planned, since Martin was immediately the subject of crude caricatures. His prominent Adam's apple and thin physique made him SMC's Mutt, of the Mutt and Jeff comic strip. Barnard still cannot look at him directly without feeling the urge to chuckle at the oversize mustache and its inadequate compensation for a hopeless lack of command in his face.

Stoole's wife was a notably different sort. Five or seven years younger and comely, with an adequate body and a generous nature, she put a new spin on "Gave at the office." She could have enhanced his academic career significantly if she had focused more on the key administrators. Barnard had never succumbed to her temptations, her unselfconscious encouragements. He was content to let Stoole's progress follow its natural course.

Barnard fixes his eyes upon Martin Stoole's receding back. His facile glibness, which is easily but unproductively penetrated, has often suggested shallowness salted with opportunism. Despite his curiosity regarding the repetitious gestures toward the material he is carrying, Barnard wishes it had been clear whether he was going to join them for lunch on Monday or not. He prefers not. He has more or less successfully dealt with the time he embarrassed his new hire in front of his own class but still does not relish his company.

* * * * *

Martin, it appeared to Barnard, had not scheduled a formal midterm exam, as Chairman Cordner set down, for the spring elective on "Central Banking and the Federal Reserve." He entered Stoole's classroom without knocking and, the students agape, berated him for the omission. The mantle of rank often facilitates such excess. Later, reinterpreting his outburst as a misdirected reaction to intense personal issues, Barnard would recall the seated students' empathetic embarrassment. He would attempt to fathom the true reason(s) why he, at one moment quietly walking the corridor scanning the semester's schedule, at the next came to be standing beside the lectern, slapping down on it, demanding corrective action. His tirade was overblown and directed at an undeniable triviality. Martin made no response, except to put out his hand to calm his fluttering lecture notes. He offered neither defense nor rebuttal, showed neither anger nor offense. He never has, as if the unfortunate event was of no concern. To be of no concern and to be forgotten are quite different eventualities. Whichever it was for Stoole, both are in dramatic contrast to Barnard's habit of nocturnal iteration.

When Barnard told Diane of it, she repeated her frequent rhetorical admonition: "Why do you let trivialities upset you so? It's not good for you." As always he had no answer. The niceties of civil discourse require that one suppresses anger. Diane believed and Barnard likewise came to recognize that he preferred to overreact immediately to the inconsequential. He thus was able to shed at least a layer of accumulated unpleasantries. The big issues went deeper, formed an intractable bolus that choked him, made him go quiet.

* * * * *

Never said a word about it. Hah.
Just Phil for lunch, please. No Stoole.

"Sal, I'm going upstairs," Barnard says. He is briefly inclined to lean over and brush Sallie's cheek with an air kiss, but refrains. "Need to freshen up. I probably snacked too much this afternoon. I'll see you in a bit. Right? We can watch a movie," he suggests.

"Oh, yes. I meant to say," Sallie says, suddenly animated, "I've a good one."

"What?" he asks, naturally curious.

"I'll tell you when I come upstairs."

He grins a goodbye to Sallie then does pretty much the same to May. As he threads his way amongst the scattered tables and chairs he is intercepted.

"Barn," then "Barnard!" repeats an insistent voice.

Barnard stops. It is several seconds before he notices the familiar musky aroma. Frank Kurchak, a crumpled up napkin in his hand, is dabbing at his lips with it as he approaches. Barnard studies his friend. He takes the neat sideburns and clear cut short hairs in Frank's ears as evidence that he, also, has been to a barber.

"You got that notice, last week, didn't you? Did you go over it?" Kurchak asks, projecting the unease that is now so prevalent, a staple of their generation. His plastic-framed glasses and broad, flat face make for an owl-like visage that suits his demonstrated commercial prowess.

"I did, Frank. It doesn't mean an awful lot," Barnard says, watching for his friend's reaction.

Frank had asked the same question two days before. Barnard's misdirection this evening is a pointless test, since Frank is accustomed to obfuscation and inconstancy. He had built a successful import business within a Chinese mercantile system rife with hidden aims, bureaucratic interference, continual renegotiations, and unabashed manipulation. Most of the early Sino-phobic Western businessmen had badly managed their perceived lucrative opportunities. Frank, on the other hand, had forged early ties by wisely being tolerant of prevarication, had learned to make his interests serve theirs and, thereby, theirs serve his. He was rewarded amply for his flexibility. Barnard sees the paunch that respectfully attests to Frank's entrepreneurial success. Also, he had been a good mentor at SMC, which had prompted Barnard to relax his bias that being book-smart was necessary for pedagogic relevance.

And all safe in dollars. Then. Snookered like the rest of us, now.

The notice in question is another invitation to relocate to an Elder Eden. The hook this time is that a prompt decision will guarantee availability of an apartment at that "nearby" facility, the one outside Silverado in the hills of Orange County on the edge of Cleveland National Forest. By intent, Elder Edens generally are located far from major metropolitan areas. The next nearest site is near Boron, west of Barstow in the desert near what used to be called Edwards Air Force Base. Its location is not as severe as is that of the Elder Eden in Searchlight, Nevada, but is equally isolated and hopelessly far from the sea. Barnard does not agree with the economic logic behind the Elder Eden program. Yet, he must accept that it is underway, that they are serious about it.

"They're trying to pump up the overall level of interest," Barnard adds slowly. He knows he is dissembling. He feels certain that request and offer soon will evolve to edict and enforcement.

5 - - SUNDAE NITE

* * * * *

Barnard had received the same invitation and embedded link. The Orange Country Elder Eden was beautifully rendered via a series of fast-paced video vignettes. It was, in his view, a too slick and overly produced presentation. Designed to be compelling without being explicit, its aim was to entice, to relieve proximate apprehensions. At his first viewing, Barnard could not decide if he were particularly discerning to have detected the factual overstatements or if the creators simply did not care that these might be detectable.

The facilities section opened with an image of a large gate across the head of a curved entry drive that led up to an attractively detailed, multistory building. A roughly parallel footpath, bordered by mulched beds of flowering plants, wound past several decorative palms. Shifting to the interior, there were full circle pans of typical apartments. The patio sequence showed a colorful beach sunset, with human forms silhouetted in the foreground. The presentation went on to cover the Elder Eden's range of activities and the highlights of the area nearby. There were clips of a group of four enjoying a casual lunch, then of a park replete with active young people mixing with the elderly. In another, a man in sunglasses and distressingly ugly shorts stood on the front step of a van, beaming out at the camera. The same man, in different attire, showed up later, amongst a diverse group of peers at a table in a room with many bookshelves. They were being addressed by another focused senior, a woman standing behind a lectern. The presentation closed with a reprise of the opening shot, a wide aspect view of the facility framed by an open gateway. There was something written over the top of the gate, imbedded in the wrought iron scrollwork, which Barnard was unable to make out. In the background, behind the voice of the unctuous narrator, was the pulse of the surf.

The incongruities were obvious. Hearing the surf while so far inland? Not likely. The sight of sand and waves from a window bathed in the glow of sunset? The facility's stated location would preclude that as well. The likelihood that his own preferences were a matter of record and that the interposed coastal vignettes were tailored to his interests, while not surprising, was disturbing.

* * * * *

"You suppose we'll have to relocate, whether we want to or not?"

Frank's question collapses Barnard's brief reverie.

"Relocation seems reasonable," his friend adds, somewhat more firmly. "Only, there's a part of it that, that, ahhh ... Well, I'm not sure."

Barnard looks off into space, then at the once instantiated football hero. "How bad could it be?" he says to him, pulling back his shoulders to convey confidence in the prospect. "It seems to be a nice place. I, uh ..."

"Did you watch the whole thing?" Frank asks, since Barnard seems unlikely to finish his thought unprompted. "Did you look at the dining room, for example? That seemed nice."

"It did. Look, Frank, we're all going to need to decide eventually. It's 'Pay now, or pay later.' Right? We're not getting younger. Or healthier."

"That's true." Frank replies evenly.

"Look at it this way," Barnard continues. "All the residents here, and at the other places, too, I'm sure, are being invited to relocate. If, I say, *if* I were going to do that, I'd want to get the best apartment, the best rooms I could. So I'd sign up soon, before they get full up. It, Relocation, might be a good deal. Seems fairly decent to me, if you can believe what they show."

Barnard tilts his head expectantly, casually examining the figure next to him. Frank has already intimated his views, to which Barnard has provided a deliberately ambiguous commentary. He is almost embarrassed to have restated a message that he finds suspect. As he waits, Barnard weighs the utility of advising Frank to review all the options. They all might eventually need the "additional oversight," he could also point out to his friend. He feels his insides tighten at the implications of that unspoken phrase and pushes it aside.

"Well, right, I'm considering it," Barnard offers less than candidly, after receiving no direct reply to his implied question. He adjusts the pocket flap of his suit jacket while he studies Frank. The truth is that he is suspicious of the Elder Eden presentation, its glib formality and carefully crafted manipulations.

"Not as many people here tonight, are there?" Frank observes.

Barnard waits, since his friend's manner forewarns that a request is coming.

"I have a VieGie booked, for Sunday, day after tomorrow," Frank starts, "with my brother and sister-in-law. Would you, ah, do you have time to walk me through it again? I don't always do it smoothly."

Barnard finds the admission inconsonant. Frank has always been an independent, adaptable sort.

"Have you set up a time?" Barnard asks.

"Is nine-ten okay?"

"That works for me. I'll meet you down there. How about a few minutes earlier. Say, nine o'clock. I'll have to get you set up – if you really need help, that is. Right?" Barnard's eyebrows are up. He leans forward, obliging Frank's assent. He sniffs. "Go easy on the aftershave, Frank," he adds.

"Overdid? I don't even smell it. Okay. Nine o'clock. Sure, right after breakfast."

Barnard passes another short sigh.

"I'm heading upstairs," he says as he waves goodbye.

Although it is a single flight up to his apartment, Barnard chooses to take the elevator. Frank walks in the opposite direction, back toward the dessert table.

Established on remote, underutilized parcels of land, Elder Edens are represented to be steeply discounted alternatives to the current, generally urban retirement centers. With their economies of scale, hence dramatically lower per unit cost, they afford the government better means with which to deal with the escalating entitlements of an aging population. In addition, the availability of VieGie for remote visitation makes their relative isolation a nonissue. Thus, as the official pronouncements state at length, Elder Edens provide practical, efficient, and fiscally prudent alternatives to the financially challenged urban retirement complexes.

The argument, much like the video presentation, fails to impress Barnard. If money is the root issue, new facilities are not going to yield significant overall savings. Costs could possibly be cut if semi-professionals took care of tasks traditionally assigned to professionals and if maintenance were whittled down to near zero. Neither is going to happen, however, and would be of scant fiscal benefit in any event.

"Enough," he says aloud upon entering the empty elevator then, "Four," after he turns about.

It will be pleasant to be with Sallie, to enjoy a snack with a glass of wine. They will lounge on the couch, drift half-awake while watching a real movie, i.e., one from the nineties or earlier, per Barnard's inclination. Perhaps, in view of the hour, they will snuggle close in bed, allow some uncomplicated offering to encourage sleep.

"Four," he hears, followed by a melodic chord.

He waits for the door to fully open before stepping out.

PART II

Bent by Time — Lebovitz 1978

CHAPTER 6

MOVIE AND A VISIT

"It is nine fourteen pee em, sixty-three degrees and clear. Hello, Pro Fessor Cord Ner."

Barnard has no need to react to the greeting. He closes his door and absent-mindedly turns the deadbolt home. Then, realizing Sallie will be arriving shortly, releases it again. He commands the Daedalus Man to mute as he strides across the thin carpet of his living room. In his bedroom he stops to rest both forearms atop of the taller of the two vaguely matched chests. He scans its surface, as if comparing it to memory. Next to a shallow tray is the hard case of his nautical binoculars. Its bottom is against the wall, the black clad relic on top. They are too bulky to take on his beach walks plus they can be unintentionally provocative, as one embarrassing incident revealed. Often he uses them to peer idly out at birds, the moon, and, yes, when particularly curious, at lit windows, always being careful to have the lights out behind him. He stares for a moment at the faded white C's painted on the edge-worn leather caps protecting the objectives then lightly fondles the coiled neck strap. He had wanted optics with a long reach and, before one distant birthday, he had dropped serial hints that Diane could not miss. The 20x pair she ordered for him was too powerful for a sea-tossed sailboat; he could not hold them steady. After several appreciative hugs and a self-conscious request for permission, he had exchanged them for the 16x50 Nikons. These serve a different need now – connection. The boat is long gone.

In addition to some coins, the tray holds a spare key and a hard candy of mysterious origin, too old to unwrap. A few small sheets of paper are trapped underneath. He tests their reality with one finger, noting that he needs to solidify the terse notes and snippets of commentary the next time he is at his desk. Professor Cordner, Retired, still prefers a notepad over his phone's diminutive QWERTY when some worthy thought does come. He maneuvers the tray to one side, then to the other, finally aligning it lengthwise with the edge of the dresser before opening the topmost drawer, which is home to an unorganized mix of small items. Several ties lie neatly twice-folded and layered, stacked handkerchiefs adjacent. An elongate open-top cardboard box holds several keys, a flat battery, an assortment of pens and mechanical pencils, a LED flashlight, and a tiny pocketknife. There are also an old ring and several pairs of cuff links. The latter are no longer needed but are memory laden, as are the few foreign coins. Collecting coins is long out of style, and he has no expectation of traveling abroad. They remain where they are out of inertia, his.

Barnard's eyes shift to a flip-top box covered in a memory invoking navy blue, soft to the touch – his Dolphins.

Long ago he had put the insignia aside for his then unknown son. When straightening up or in idle remembrance, he would often take the pin out into the light, bounce it in his hand. Memories would come, as they do now. Decisions and their consequences would be reviewed as, briefly, they are now. Kyle had never indicated an interest in its history, never revealed any curiosity about the many months that Barnard spent submerged in calm waters, hundreds of feet below an otherwise often turbulent sea, sharing life in a slender tube. Therefore, he knows little of his father, just as Barnard knows little of his, but for far less valid reasons. He will feel regret precisely when it is too late, when the broaching mammals are no longer evocative, when Barnard's Biograph and VieGie residuals are all that remain for him or his own children to query. Curiosity, the same as thrift and prudence, is often idiosyncratic, is often more latent than satisfied.

Barnard's clearest impressions of his own father, who was gone before he was eight, are of wind and noise in a fast car, of the ice cream that seemed to be his reward for not letting on to Mother how far and how fast they had gone. He can draw upon only scanty memories. He wishes he could have accrued more. Kyle, on the other hand, consistently has shown little or no inclination to inquire. Perhaps this is because there is little worth exploring. There are only the tediously ordinary intermediates, from unguided beginning and vacillation, through persistence rewarded with a long, unremarkable life. Shortly after he retired, with Diane gone, the time thereby ripe for reflection, Barnard had acquiesced to compiling his Biograph, just in case.

"Pfffhhhhhhhh," he exhales, as if to dispel the memories, to extinguish them like a candle.

Barnard closes the drawer, pushing firmly against the bind midway along the warped wooden guides. He takes a step back and turns about to look left then right, slowly surveying his bedroom: the neatly made-up bed, the side tables, and the chair adjacent to the closet. The nightstand lamps are providing the only illumination. The shade on his side is canted, from his having bumped it when reaching for his glass of water during the night. He gives it a disinterested glance before turning back to the dresser, this time arching forward toward the mirror above. He sees someone familiar and imagines how its thin lips would appear if pursed. They respond exactly so. The congruence of will, sight, and kinesthesia is sufficient proof that it is he. He notices, however, that the visage is not pleasantly neutral. It does not present the relaxed look of one adaptive and content, as he would prefer. Rather, it shows a hint of apprehension. It presents evidence, which he has tried to extirpate from his unguarded repertoire through mirrored practice, of being aware of unwelcome realities. He attempts to reform his features into that of welcoming engagement. This overlain configuration melts away too quickly; Barnard is a work in progress. When one dies the intentional persona will likewise fade. Then, they say, the face will reveal the true inner person.

And what face will that be?

He glances out from his bedroom, down the short hallway. The serviceable living room lounge chair, his couch and its two tables – one to the side, one in front – cannot be seen, only the side of his bookcase, a slantwise view of its minimal contents, and his entry. The perspective seems to foreshorten his space, make of it an objectivist painter's preliminary sketch, a still life of the inconsequential.

Barnard turns back to his mirror, reviewing what he might have missed by skipping dinner at Sunset Vistas tonight. It would not have been much, he imagines: breaded and overdone fish; a side of something green – broccoli or long beans or limpid spinach; a salad of wilted leaves under sliced tomato and/or onion. Barnard has taken to declining the faintly yellow dressing regularly set down for them. Its color, its faint odor of vinegar and cheese, the thin, semicircular handle of the dull metal server, are unpleasantly reminiscent of the college lunchroom from another life. He scans his reflection to determine how acute is his need for a shave. Marginal, he judges.

"Funeral. Old Country snacks," he says aloud, watching the way his mouth shapes the words.

Right. Should have taken a walk.

Removing the coins from the small tray, he arranges them into a single irregular stack according to denomination: two gold toned fivers – erroneously termed Bronzies; several dollar coins; a quarter. Then he adds the dimes, with nickels atop. He rarely has occasion to carry or receive coins now. His electronic account suffices. When in middle school it had been fun for Barnard to collect change, to sort them by date and mint mark, especially the stately Walking Liberty silver half-dollars. These, even then only rarely gleaned from circulation, were spent much too soon, well before their value rose to many times their denomination.

Pennies? Gone. And Penny books, also. Dark blue, navy blue. Where?

Hardly a stack, Barnard decides. And unstable obviously, since nickels are larger than dimes. He restarts, arranging the coins anew, this time in a graduated spire from large to small. He starts to push that over as well then, thinking of their small community's situation, flexes his finger back. What he had superficially shared with Frank is not far from the truth. If Sunset Vistas were to default, relocating would be a necessity not a choice. Barnard does not want to believe it could come to that.

He realigns the three gatefold picture frames then folds his virtually pristine handkerchief into a neat square and rests his wallet upon it. He stares for a moment at his cellphone before adding it to the formation. His mind far from the present, he languidly returns his gaze to the mirror, tightening his lips into thin lines then pulling the upper one up. It is semi-intentionally much like Bogart's distinctive, toothy grimace. Barnard cannot help but try to recall when, if at all, that showed up in *Casablanca*.

With Ilsa? After? Leaning against the bar? Talking with ... with ... the fat man. Farr ... Ferrari?

Enough. To hell with all that.

He pulls his shoulders back and marches to the living room, waving one hand over the switch on the wall as he exits. The lamps on both sides of the bed slowly dim. Standing before the living room window, he sees that all traces of the sun's red demise have faded. Bluish white light, from the diminishing sequence of light standards along the curb, outlines the palms in the near distance. Their umbrellas of drooping fronds are still. There is no shore breeze to disturb them this night.

Barnard runs the backs of his fingers up his neck. This confirms that he has no need to shave but does elicit the inclination to wash up before Sallie arrives. He retreats from the window and goes back along the short hallway to his bathroom. The recessed ceiling lights come on as he crosses the threshold, which they do not do between 11 p.m. and 7 a.m. – a rare example of obeisance to his preferences. No sense shocking themselves fully awake or disturbing each other when they get up to use the lav. A face in a different mirror, the one over the vanity sink surrounded by faux stone, greets him. It is the same face, but older, with creases that are accentuated by the strong down light. Yes, those are the same teeth, the same eyes and nose. He is still Barnard Cordner, Retired. He glances up, reviewing, yet again, his tentative plan to ask them to insert a diffuser or baffle to cut the glare. But would he still have good light for shaving? he wonders as he realigns his pill containers. He puts both hands aside the sink and leans in toward the mirror.

Non-prescrip, mostly. Lucky. Unrestricted.
Better wash up.

If the first of the recent twin market disasters had come before he retired, when he was active and had the means as well as opportunity to extract some advantage, Barnard might have welcomed a major downturn. He had been projecting that eventuality. His shadow portfolio of candidate basic industry, technology, and consumer stocks was compiled and ready. While he did not explore the myriad ways of potentially choosing or allocating poorly, he long foresaw that potentiality, that likelihood, in fact. He could have bought at an apparent bottom only to find there were further bottoms to be tested. He was well aware of and had taught of how retail investors are generally poor traders. Being a contrarian is a step up but, since the reasons for sharply lower valuations can be as valid as for higher, that also is not very profitable for a nonprofessional. It is experience and patience, not adage or simple rule, that foster successful investing. Recognizing that his academic understanding of economics was less help than hindrance in managing his financial affairs, he had left his retirement assets for others to watch over. A tax advantaged, diversified portfolio of funds, with a bond component that grew larger as he aged, was wiser than speculation. The current miasma has proven to be an exception.

Exigent circumstances.

Barnard applies a dollop of paste to his toothbrush and runs a brief stream of water to dampen it. Deliberately, he works it over his teeth and stares into the sink basin. He pokes it under the spout to restart the flow for his rinse, then replaces it, bristles upright, in the narrow cylinder on the left, aft of the pill containers. Sallie's few items are on the right. Thoroughly preoccupied, Barnard glances at his naked wrist; he has not worn a watch for over a decade. He shakes his head, showing a hint of annoyance before looking at the barely discernable numerals on the wall switch.

Old habits reemerging.
Where's Sallie?

There is so much to worry about, so many decisions to make – for each of them. He leans into the mirror again. His two hands cradle then lightly pinch downward on his chin. Noting the prayerful appearance of the gesture, he quickly lowers his hands when he hears his entry door close.

Should shave?

"That you, Sal?" he calls out

He had not heard her key in the lock, possibly because the water was running. Probably. He deliberately puts on a less dour visage.

"Yes."

"Hi, Hon," he says before reaching the living room. "What's going on with May?" he asks as he approaches.

"Nothing really. Oh, or everything, I guess." Sallie carries a white paper bag. Observing Barnard's steady, questioning stare, she adds, "I brought us a treat."

"And what would that be?" he asks, trying to reinforce his altered mood, hoping to expand and make evident his sincere curiosity.

He kisses her lightly on the cheek. His arm slides from her shoulders down over the curve of her slacks, which he pats affectionately. He steps back and follows her into the kitchen, where he eyes the promising white sack she places on the table. Inside he finds two squares of richly iced cake. One is chocolate topped in white, with a dollop of chocolate centered on top. The other is its complement – white cake covered with dark brown and surmounted with a kiss of white. They form a cubist portrait that stares back and tempts him. He pushes his face in close and sniffs.

"Where did these come from?" he inquires. "Not here, that's obvious." He is fully real again, relaxed, and smiling sincerely. He gives her another, lightly anticipatory kiss. "I can taste, uh, taste ..." he starts before licking his lips. "Chocolate! You're bad."

"Don't be silly. It's coffee." Sallie reaches over to close up the bag. "And that was hours ago. When I was out with May. I thought to buy a couple of pieces for tonight and put them in the 'fridge. I hope they're not too cold now." She reopens the sack.

"I'll heat them up, Hon, if you want. You, too, while I'm at it." Barnard leans against her.

"Don't they look good?" she asks. "Don't touch the chocolate one. That's mine," she teases back.

"Don't worry. Not a chance. Not this late." After a pause he adds, "You're sweet."

"I'm joking, Babe," she replies. "They're both for us. We'll share, if a little chocolate'll be okay for you."

Barnard shakes his head to reaffirm the well-established No. They look at each other fondly. What they have is neither too little nor too late. It works for them.

"You've got white on your chin. Have you been snacking already?" she accuses. In fact, she recognizes what the bleb truly is.

"No. I was, uh, washing my, uh ..." he starts to say. Guided by Sallie's eyes, he samples the offending blotch with his finger.

"Toothpaste?" The crease between his eyes signals recognition of his error. "Toothpaste, right," he affirms. "Funny. I meant to wash up. Hah." He pivots about. "I've been going over this Elder Eden business," he lies over his shoulder as he walks briskly back to the bathroom. "It's a pain. It's all truly becoming a big pain."

The lights pop on again and there is a splash of water as Barnard washes his face and hands. He takes a firm swipe at his chin, crinkling it up and forward to be sure.

"Anything particular you want to watch, Babe?" comes down the hall while Barnard is draping the towel over the bar under the sink. Sallie's "Babe" is more endearing than is "Barn" or "Barnard" and does not simply echo his own familiarities with her.

"Yes. I have a good idea," she adds, continuing to pointedly avoid addressing his last comment. "We overheard, May and I overheard someone talking about *Hot and Cold*, while we were having a coffee. They kept interrupting each other, as if it was an exciting movie. So I thought we'd try to find it."

"Who were talking, Sal?"

"Oh, some people at the next table."

"And where was this?" he inquires.

"A few blocks off Wilshire. No, an ordinary coffee place, Babe. Don't start planning any new excursions. That movie might be a nice change from that old stuff you like to watch." She has anticipated his inclinations; two digs in seven seconds. She looks at him expectantly, her raised eyebrows pleasantly soliciting a reply.

Barnard considers whether to feign deflecting an insult.

"We should just put these on a plate," he suggests instead, "to let them warm up. Why do you say it's recent? The movie, I mean," he then asks. He puts his nose close to the chilled sack and sniffs lightly. "It could be an old classic."

"Well, I've never heard of it. It was foreign, from what they were saying." She stands by the kitchen table. "A plate? Yes. Get two, Babe, please. Take them out and –"

"I'm kind of tired, Sal," he interrupts. "Why don't we pull down the shades, get changed, and nibble in bed?" Barnard assumes a stiff demeanor, puts his nose up. He raises one eyebrow. "Heh, heh, heh," he breathes, to recreate a tux-draped lech from some old comedy.

"And what, my dear, is to become of our sweet treats?" she coos with mild mock concern.

Barnard laughs at how she has slipped so sweetly into character with him. He passes his hand lightly along her arm before going to the window. The illuminated, untroubled palm fronds are progressively cut off from view as he lowers the shade. He takes one last look down to the empty street before it, too, disappears behind the opaque ivory slats.

"Well, what I think is that we should have these right here and get ready for bed after," Sallie proposes. "We can watch the movie all snuggled up. Doesn't that sound better?" Sallie asks.

"It does. Right. Sounds good," he replies agreeably. It has been a long time since he snacked in bed, once a delicious idea to him but foreign to Sallie.

Milk? Yes, milk. Calcium.

"'It all starts with milk.'"

"What are you mumbling, Babe?"

"I was remembering what my grandfather used to say, at the dairy." He sets down a pair of salad plates by the white sack. "I'm going to have a glass of milk. You want some?"

"Just water, please. I've been thirsty all day," Sallie sighs as she carefully lifts out the decorated pieces of cake.

"Probably because you and May talk so much," Barnard observes. He puts down forks and napkins then pours milk for himself and water from the fridge for Sallie, wiping the bottom of each tumbler before setting them before their respective places.

"Should we share, Babe?"

"No, that's okay. You have the chocolate one. We can trade some of the icing," Barnard is quick to propose. "Did you go far?" he asks between bites.

"No. Just up to Third Street. This is good, isn't it?... May ordered a second piece for us to share," Sallie then confesses with downcast eyes, shamming guilt. Barnard points to the supplementary implicating evidence before them. "Yes. We're bad, too," she laughs.

They are adjacent at the table and eating cake. Two sinners simply enjoying. That is how it has been from the start between them – making small moments count.

* * * * *

Sallie's late husband made a decent living, took care of their finances, and took care of her. He usually did the cooking or brought prepared meals home. They did what middle class, childless couples generally did. There were local entertainments, social events, and quiet evenings at home on the weekend. There were restaurants, Broadway shows at Santa Monica Civic, Century City, or even downtown L.A. There were carefully selected, relaxing cruises, which alternated with arduous treks, such as guided safaris through African game preserves and by themselves in the Dolomites. They had a list of adventures to complete before their stamina gave out. Then, without realizing its decline, most were beyond them. They never walked the spine of the Pyrenees, never took the stairs, even down, at Eiffel's tower, never immersed themselves in the rigorous religiosity of Indian coastal temples. Various hobbies and community activities were compelling substitutes, but their interest in each would quickly wane. Bridge and gourmet clubs, as well as the personalities of some of the participants, likewise had grown wearisome by the time they began to take retirement seriously. They had tentatively explored moving further up the coast or to Arrowhead, until they saw the prices. Salsa dancing was the last new thing they tried before, in an instant, he was gone and she was alone.

Their modest aspirations matched modest means. Planning had not been complicated by overambitious intentions. When distant future unexpectedly became proximate reality, there was loneliness and too much quiet. Sallie felt the unfamiliar, unsettling press of solitary decisions. In addition, there came that part of the economic cycle that few prepare for, the downside. It did not take long for Sallie to determine that the house and their investments for retirement were worth substantially less than she had anticipated. She, her entire generation, had enjoyed a unique era of steadily increasing assets, which had come to an end. Alone, she spent much time regretting not having paid attention to their financial situation, what it truly was and how to manage it. Until singularity was thrust upon her, Sallie had not realized how profoundly she had been sheltered by a responsible yet unsharing spouse.

Pregnant Fridays and silent Saturdays were the hardest part of each lonely week for her. Friends, who had dealt with similar separations, managed by closing ranks with family. Sallie had no one. On quiet afternoons, she often envied those who had invested a part, even if a begrudged minor part, of themselves in children. Buying into Sunset Vistas provided her with what she needed, with more, in fact, than she had reason to expect.

* * * * *

Fully immersed in the present, they savor the last few bites of their cake. Sallie looks aside at the silent Barnard then wipes her lips and leans lightly into him before replacing the napkin on her lap.

"And where have you gone off to, Babe? You were fun and games a few minutes ago. Sad about your auntie?"

Barnard takes a long breath and holds it before responding.

"No. Just about things in general. You've been sort of quiet yourself. May said she had a question? Were you two talking about Relocation, about the Elder Edens? That doesn't make for a happy mood."

"Yes. We were. It's that, and it's Chile and her kids that's bothering her. She wants to hear what you have to say. Let's not go over that again, though. Not tonight. Okay? I'll clear and we can watch that movie in bed." Sallie is already collecting the empty plates and putting them in the dishwasher. "I'll just rinse them off. Still stuff from yesterday?" she notes. "I'll finish loading and you can run it in the morning. I'd prefer a light breaky tomorrow. Just something up here, maybe. Okay?"

She opens the cabinet under the sink for the detergent and fills the dispenser.

"I'll set it to start tomorrow," Barnard says as he moves next to her. "For after we go down for breakfast." He rinses the glasses and places them in the nearly full top rack. "Let's go down say at, uh, at eight. I'm supposed to meet Frank at the big VieGie at nine. No, wait. That's on Sunday. I don't have anything going on tomorrow."

Somewhat preoccupied, he has not fully parsed what Sallie said. She pays no mind. She is as accustomed to Barnard's mishearing as she is to his need for subtitles. Both, however, hear the rustle at the kitchen window. His apartment is on the fourth floor, so it cannot be anything ominous.

"What is it?" she asks, as Barnard starts toward the sound.

"A pigeon," he tells her. "Started a few days ago, flying up here at dusk. It'll go quiet soon. I hope to hell it hasn't laid an egg."

Barnard presses down on one of the shade's slats. Sallie stands quietly beside him. They study the source of the repeated "Coo ah. Coo coo." A glistening bead of an eye peers at them in return.

"Odd eyes," he says. "All black. That must be because it's dark out."

"Aren't their eyes usually sort of orangy?"

"Right. Usually," he affirms flatly. After a pause he is more animated. "I'm glad it hasn't tried to fly in. I'd have to, uh ..." he starts to say as he moves away from the window. "As long as it doesn't poop on the sill."

Barnard gestures to lower the kitchen light. A few steps behind, he follows Sallie toward his bedroom.

"Why do they call it – what we don't do – 'the birds and the bees'?" Barnard asks, trying to sound serious.

He grabs at her behind and quick steps into the bathroom ahead of her. He laughs, satisfied that the gesture was sufficient.

"Be quick," she says and stops.

"I will. Quick as sex."

Tiny dribbles of foam slip from the corner of his mouth as Barnard brushes, up down, across and far back, for the second time this evening. He repeats the motions to excess. Sallie enters and interrupts Barnard's reverie at the sink. He gives her a playful hip bump, to which she replies in kind.

"Hello there, Sal. Tired?"

"A bit," she says. "I'm looking forward to that movie. I'd like something different tonight."

"Great idea! How about we fool around, Sal? Do the old in-an'-out, in-an'-out? I can make it different." Toothbrush askew between teeth and cheek, Barnard pulls out the front of her blouse. His laugh anticipates hers. She slaps him lightly on the hand. After a final rinse, Barnard replaces his toothbrush, its damp bristles upright. Lingering, he stares into the mirror.

"What are you looking at, Babe?"

"An old man, with stubble," he replies, suddenly overtly morose.

"That's all right."

"Pffhhhhhh," Barnard emits. "That's not what you're supposed to say."

Sallie pushes him out then closes the door.

"Actually, inside I'm a young, handsome prince," he calls, his mood suddenly swinging back. "Kiss me and see."

Padding along in short steps, he accelerates into the bedroom. One of Barnard's three identical sets of pajamas always hangs on a hook behind the door. If it were not for Sallie's occasional reminders, he might slip back into going several weeks before throwing his into the hamper. After changing he folds down the coverlet and flops into bed, tightening his legs to the edge of cramping before pulling the light blanket briefly over his mouth to cover a long yawn. Remote extended, covers tight to his chin, he awakens Daedalus.

"It is ten oh nine pee em, fifty-eight degrees and clear. Hello, Pro Fessor Cord Ner ... and Guest," issues from the screen, in its familiar flat tone with the idiosyncratic syllabic breaks. "Would you care to join in a short evening Pause for Prayer?"

These spiritual homilies have much in common with the sermonettes that used to conclude each broadcast day when he was in grade school and many local stations were unaffiliated. Except, they now are personalized, interactive, and insistent, as are other Internet conditionals. If no selected program is in progress, they pop up every odd hour in addition to noon and midnight. While occasionally cajoled, as are most, Barnard has never received any officious reprimand for his refusal to participate.

"Why the hell does it have to watch everything we do?" he yells out to Sallie who is still making ready in the bathroom.

Barnard speaks as if he truly wants to know. He does not. Daedalus is not a real person, after all. While superficially intrusive, an interactive and always alert Internet terminal does provide conveniences. He does not have to verify his status in the morning and at bedtime, as his mother had to do. It adds a measure of security and he can anticipate a prompt response to any medical emergency.

"Subtitles okay?" Barnard proposes through a sincere smile as Sallie pads up demurely.

She slips into bed without answering, touching his cheek with her lips before pulling her share of the blanket also up tight and cozy.

"I'm cold," she says, although Barnard can feel her warmth. "My tootsies."

"I'll make us a toe sandwich," he offers. She puts her feet side by side so that he can capture them between his. He squeezes her toes lightly between instep and arch. "Is this good?"

"Yes. Good, always," she replies.

She reaches behind to adjust her pillow and pointedly looks at the screen then at Barnard. She has already suggested what they should watch. Barnard has it highlighted on the on-screen menu. *Warm, Hot, Cold* is the actual name of the movie. At least, that is the superimposed translation of the calligraphic script that is flowing across the screen.

"Right," Barnard says to her. "Video. Play," he then commands, in that special voice.

The cyclic repetition of graceful script and unfamiliar Eastern music fades. The trailer leads Barnard to suspect the film is not what Sallie imagined. Close to each other and leg over leg, he fears that a Mideast action tale with a steamy subplot of sex is a poor choice for this night. After the opening credits, translated into jiggling, limpid overlays of text, the story begins.

Warm, Hot, Cold is set during one of the Muslim wars. They never make clear which one. There are, as usual for the genre, nuggets of instruction that suggest that it is the first, that vicious, circumscribed preamble to the wider internecine battles still to come. The action centers on a pretty young woman whose innocence does not save her from being caught up in the war. Following a sweep of her settlement, she is escorted to an Education Community, a tightly secured compound far, in distance as well as in sectarian orientation, from the family-centered comfort of her home village. She has no choice but to accept her loss, the imposed duties, and the indoctrination. There are scenes of her being firmly guided into traditional dress, behavior, and respect. There are sharply portrayed images of initial resistance and rebellion, of her failed attempts at negotiation and compromise, of her gradual assumption of the subdued pose of a convert. She remains lovely throughout, as the film spends much time exploring.

After one particular instructional session, she responds to the attentions of a handsome local. Images of a past life – hers – and adventures – his – overlay scanning shots of their engaged faces. He befriends her and soon is her ardent, if inconstant, lover. The relationship provides her with ancillary benefits: less supervision, marginally enhanced status, special tidbits, and the opportunity to leave the dull impositions of a segregated dormitory. She moves from resignation to a new cycle of action and reward. It is a fertile plot and they do a nice job of bringing it forward.

Several scenes are in the cooking area of his quarters, with her engrossed in preparing one traditional dish or another for the two of them. Her obvious displeasure with the Spartan facilities reveals her perception of reduced status and suggested that she had been the jewel of a well-to-do family, albeit of a minority sect. Through the way her eyes follow and her fingers subtly mimic those of a man intently stroking the raab, it is made evident that she probably has been taught to play. Even sitting, the lithe heroine moves well to the exotic, almost sensual rhythm of the drumming. Other vignettes provide evidence of her education, of her exposure to more than the domestic arts, all of which she has had to suppress.

An inner climactic scene follows from her lover bringing in a bloody lamb leg and tossing it onto his rough worktable. She had requested it for a special dish, something locally unfamiliar – moarraq. Standing before the table, her spirited dissection with a narrow-bladed knife is a deliberately focused interlude, one using repeated, tight shots of red meat, beige bone, and glistening white tendon. She becomes increasingly ferocious with the short, narrow boning knife.

It seems too long and detailed a scene to be a cooking demonstration, even if tantalizingly ethnic, Barnard concludes.

Must be going somewhere with this.

Sallie tightens and shrinks back under the covers from the sound as well as the 3-D gore of it. He notices her grimace, thinks to ask if he should turn it off. Suddenly the carving and cubing are over. Her tension eases as the young woman proceeds to slice onions, which sizzle noisily when dropped into a pan of hot oil. Her protector adjusts his head scarf and retreats a few paces. After a few lines of dialogue he leaves. Alone in the kitchen, she smashes cloves of garlic with the flat of a bigger knife and adds these, plus a grab of salt and a double grab of an orange spice, probably cayenne pepper, to the smoking oil. She throws in the cubes of bleeding meat, which she stirs with a wooden flat. Giving it a long look, she adds another palm-full of the pepper.

Christ. Hot. Really, really hot.

The audio focuses on the meat's sizzling and popping. They can almost smell the lamb, the oil, the onions, garlic, and spice. The young woman adds enough boiling water from another pot to cover the meat. She lids the pan and reduces the flow of propane. From the bowl of speckled olives nearby she selects one to try before dumping them in with the meat. The rice, in another metal pot, gets several passes of rinse water before being covered and put on the burner. The scene concludes with a slow dolly back. She is motionless and facing away, hands backward upon her generous hips.

How so many pots?

The dish has looked so wonderfully exotic that suddenly, as if on cue, Sallie and Barnard snap their heads toward each other

"We should try that!" they say in spontaneous unison.

They laugh and shift about, leaning in close to one another. His hand slides down her arm then over her legs. The fabric feels smoother than skin. Her head tilts toward his.

"I'll write in for the recipe," Barnard jokes. "Smell definitely needs to be added to the 'net," he adds, in mock seriousness. "I'll bet they're working on it."

"Sshush."

The film never shows the meal being enjoyed. With the plot well underway, its pace slows. The presumptive heroine learns much in addition to cooking: Mideast aspirations, Zionist scum, the decadent West, how to make clothes and care for children. She dutifully studies alternative interpretations of the Koran. It is a different life for the once sheltered girl and its detailed portrayal is not particularly entertaining. Such is her metamorphosis from a privileged to an observant woman, the film obviously is striving to convey. It is natural for Barnard and Sallie to ponder what might be her eventual dramatic crisis, what might be the crux of the story. Barnard, in particular, is beginning to lose contact with the decelerating plot. In an undoubtedly meant-to-be-funny scene, which reawakens his interest, she fires a fully automatic rifle and terrifies nearby trainees when she loses control of it. There are a few explicit scenes of love making, during which Barnard experiences a private tingle. He glances sideways at Sallie, moving only his eyes. She is quiet.

The girl's protector is handsome and acceptable until he allows a close friend in to enjoy her as well. She cannot object. She owes much of her present and possibly extraordinary satisfaction to him. Then he has her undress slowly before favored comrades and young jihadis, receiving small cash payments from the former and providing anticipation of future seventy-two to the latter. She cannot object to this either. She is trapped.

Barnard's tingle increases as he expands.

Then the handsome protector begins offering her to anyone willing to pay, or to those to whom he owes a debt or from whom he desires a special favor. The on-screen pair argue over the lover's demands and, nearly hour into the story, Barnard's eyelids feel supremely heavy, such that he misses a "page or two" of development. The climax comes, so to speak, one moonlit night as her lover strains on top of her and is nearing his finish. It is a fully developed sequence, shown in multi-angle, merging overlay close-ups with enhanced sound, deservedly sans translation, which bring Barnard to full awake and familiar stiffness. The girl lifts her arms and puts them aside her head. Reaching behind, she grasps the handle of the previously introduced boning knife, now stashed under their pillow. She opens her legs and lifts her hips upward in a lordotic pose that, her eyes reveal, is not from passion but to evoke her lover's sense of mastery. Slowly she slides the sharp blade down her belly. Then, with a sharp thrust she severs the fleshy bond between them. As he leaps and clutches and screams, she reverses her grip on the knife to plunge the narrow blade slantwise into her chest.

Uhhh! Debonered by a boner. Hah!

She has placed the tool, now a weapon, exactly right. The hilt is firm against the skin of her naked and still chest. Both music and camera work have made clear that this final act had been carefully planned. Barnard's tumescence fades rapidly, as if he has been cast into a cold shower. He grunts lightly.

Well done.

The rest is wrap-up and credits. Barnard reviews the dramatized rewards, the patient justifications of the story, the stark portrayals. He shifts position to gauge Sallie's reaction.

"This isn't the kind of stuff you like, is it? Did you mean to pick this one?" he asks in a voice made husky by trying to speak softly.

She remains quiet and unmoving beside him. Barnard should have noticed her slow, regular respiration.

Asleep. How long? Just as well. Too gory.

Her exhalation thrills the hairs on the back of his hand as he smooths the blanket beneath her chin. He wishes they had picked a different film. He remains awake for a while, dwelling upon the film's cruel plot and its onion-like layers of messages. He notices that he feels no frontal eyestrain headache, as he often gets when watching in bed.

"Interesting," he offers to no one, teetering on the edge of sleep. "Not dizzy. Maybe the sex scene."

Sallie stirs, rolls on her side. Barnard squinches down and rubs both hands along his inner thighs. She, he, the night are quiet.

Everything you want's on net. Good movie. Too bad she ...

His thoughts turn to the weekend, what he might do, as if it matters. Tomorrow he might take a stroll on the Venice Boardwalk and find a snack; it should be lively on a sunny Saturday. Perhaps he can coax Sallie into going with him. On Sunday morning he will help Frank with his VieGie, as planned. The timing is good. He has one of his own, with Kyle and the kids, scheduled for later that same day. Sunday is always a good day for VieGie.

Barnard shifts, trying to find a cooler spot, and lifts the light blanket from his chest. He cannot control the intrusive flashes of reality that resonate with unwelcome concerns.

Formula films. Right. Somber symphonies of generated graphics and high-testosterone pop action. Their collective style, not ours.

His eyes dry from staring, he blinks slowly and wills the quiet darkness to envelope him, to subdue him.

It seems only an instant before Barnard awakens with an urge to use the lav. The TV's time bar shows 2:47 a.m. He gets up slowly, bending sideways so as not to disturb Sallie's half of the blanket. He exercises the same care upon his return, sitting carefully on the bed then unfolding slowly backwards after flipping his pillow over onto a new, cool side. Through lowered lids he stares at the red dot at the bottom of the screen. It presents a faint aura around an irregular asterism due to the expected accretion of corneal imperfections. Cataract surgery is always an option. Unfortunately, the longer he waits, the less likely he will live long enough to be scheduled for it. If he were still on the faculty, then a year might be as long as he would have to wait. Two, or even three, is the norm now, for retires such as he.

Joys of aging.

He closes one eye and fixates on the fractionated bleb of red. The outline of the doorway beyond fades into a faintly speckled grayness. Forcing his gaze to remain steady, he waits for the red target likewise to disappear. It dims, wavers, and shrinks but persists. He cannot suppress the frustrating saccade that brings it all back.

6 - - MOVIE AND A VISIT

* * * * *

Barnard used to amuse himself that way when a sleep resistant child. Shortly after Father died, he had been staring at the faint patch of greenish-blue light on the wall past the foot of his bed, intent on keeping it steady, when it abruptly disappeared. A tingle from the electric shock of believing he had gone blind, he shifted his gaze. With that, the phosphorescence and the suppressed world around it magically reappeared. His vision was saved! The cold thrill eased. He sat up to verify the window's faint outline, the yellow glow from the hallway seeping under his door. Thus reassured, he repeated the process several times before heavy sleep overpowered his curiosity.

It was thereafter a pleasant game when restless at night. After he perceived the possibility of doing irreparable harm to his eyes, like that from staring down at the tip of his nose, of which Grandma Jane-Jane had often warned him, he indulged less frequently. Later learning, in school, of the underlying biology of reversible sensory fatigue and photopigment bleaching, which accounted for the transient loss of vision, took away the mystery of the illusion. In any case, the game was soon superseded by the other amusements that are part of adolescence.

* * * * *

The Internet terminal is never really off, even when the screen is blank. It remains on perpetual standby, every ready and taking in. Barnard's visual game with the red dot, its sign of life, has, therefore, been resurrected in recent years. Unfortunately, it is not a reliable meditative talisman, not always an effective aid in his striving to suppress circular ruminations. Tonight it takes some time before he slips back into sleep.

* * * * *

Saturday was uneventful, simply another pleasant coastal day to savor and enjoy.

* * * * *

Frank is not there when Barnard enters the empty VieGie room at a few minutes after nine on Sunday morning.

6 - - MOVIE AND A VISIT

"Schedule," he says to the screen. It dutifully provides the Sunday morning line-up. A few lines down he spots "Rumbauer Mary :: 8:30a >> 9:00a." After that is "Kurchak Frank :: 9:15a >> 10:00a."

"Pfffhhhhhh."

She must've finished early.
Nine ten, he said. No matter.

Barnard surveys the room for a remote control, just in case. He spots one on the stand between the pair of large armchairs and waits patiently. It is not long before Frank joins him.

"Hi, Barn. Kept you waiting? Sorry."

"No problem," Barnard replies. "I just got here. You have everything you need?" He can barely detect the scent of Frank's recent shave.

He minded.

Frank gives him a perfunctory nod. Without waiting to be prompted, he picks up the controller and taps a sequence of keys. Firmly, without hesitation, he commands the screen: "Kurchak. Frank. Local. Start." He drops the remote back in place and takes one of the heavy chairs, motioning the surprised Barnard toward the other. A couple, not old but beyond middle age, appear before them.

"It's my younger brother, Karl, and Yelena," Frank informs him peripherally.

"Hello there," and, "Hello, Frank," they say, almost in unison.

"You're looking good, you two. Thanks for coming by," Frank replies. His casual greeting unwittingly attests to the veracity of the curved and exceedingly crisp three-dimensional imagery system. Detecting that they have noticed Barnard, he turns partially in his friend's direction. "This is Professor Cordner, from the University, where I taught," he clarifies.

University? Taught? No matter.

"Glad to meet you, Professor," Karl offers. He rises slightly, sticking out his free hand, as if he could physically reach Barnard, and twitching it up and down a few times.

"I am pleased to meet you," Yelena adds pleasantly.

Barnard raises one hand in reply. He smiles but feels slightly out of place.

Frank is quickly engaged with his brother and sister-in-law. Their conversation well underway, there is no point to Barnard remaining. When he starts up, however, he feels Frank's light touch on his wrist. Fine, Barnard decides; he has no plans, no agenda. He settles down slowly using his arms then reviews his Saturday walk along the Third Street Promenade in central Santa Monica with Sallie and their sketchy dissection of Friday night's movie. She remembered only the start of it, up through the kitchen episode. Once he filled her in with the details of the remainder, she indicated she was glad she missed it.

"I am too," he recalls having replied to her admission. "That you missed it, I mean. Too graphic. At the end, anyway. But it had me hooked, so I watched it through. No," he had added, "it's not worth watching again. Well, maybe for the cooking scene. Right?"

He next revisits this morning's conversation over a quick breakfast, quickly losing focus and seeking other topics to dwell upon. His attention continuing to drift, Barnard assesses the life-sized pair in front of him. Yelena appears to be a pleasant, if overweight lady wearing a flowery print dress with padded shoulders – very East European, East German possibly. Her hair is fixed in a pair of tight braids wound tight to her head. Barnard's eyes shift laterally. Karl looks as if he could benefit by eating less as well. He lacks Frank's full paunch, but it is on the way. His skin is lightly flushed, which suggests that they had anticipated that Frank would be early and rushed to make the session.

"If it's all the same with you, Frank," Barnard hears Karl saying, "I'll figure it out. I'd rather not go over the details again. I'm fed up with the situation. And with all of them, too, if you don't mind."

Not particularly interested in what he might have missed, Barnard slips easily back into his own thoughts. He focuses on today's lunch options, whether he would rather go out or eat here with Sallie, as yesterday. It was not a very good meal, he reflects, although May had left none on her plate.

They should make better meals on weekends, at least.
I need to be doing more, too.

He pinches the flesh on the back of one hand and pulls it up. His skin is so dry and so wrinkled.

Sal's right. Need lotion.

While Barnard feels he must appear attentive as they talk, he makes a conscientious effort to not listen. He cannot fathom why Frank has him stay and vainly tries to recall relevant facts instead. He knows Frank has no children. The manner and appearance of the visiting pair, the lack of any familial content in the brief snatches of conversation that Barnard attended to at the onset, suggest that they are childless as well. He would be hard pressed to justify that determination. If true, however, it would mean that

Frank and his brother represent the end of their line. Karl and Yelena have always resided abroad, Barnard additionally recalls Frank telling him. Also, he vaguely remembers Frank going for visits. Karl is in sales or light manufacturing, and most likely did business with his brother, in China. Frank had mentioned that they worked together. Right, it must have been a family business, Barnard thinks. Or, in light of Karl's expression of frustration, it still is.

Barnard studies the VieGie display, evaluating the extent of fraternal resemblance. It is obvious that neither Frank nor Karl would blend easily into a room of Chinese businessmen. When Sunset Vistas is mentioned, Barnard, his attention focused elsewhere, does not capture the full sense or context of the comment. The data line shows it is already 9:32. He cants his head, hoping for a pop of synovial cavitation, which often seems to ease the stiffness in his neck. Even if myth, it is worth repeating. Craack, he finally hears from within. After several moments Barnard once again braces forward on the arms of his chair to stand. Frank pokes the side of Barnard's leg then pats the air behind the table between them, another signal to remain. Barnard leans back, sighs softly, and crosses his arms, his hands tucked under. He gradually leaves his own thoughts and turns his attention to his friend's conversation.

"Yes. We've received several emails from the Elder Eden people," Karl is saying. "It's pretty nice, I'd say. Have you decided to take their suggestion to relocate?"

Frank presses Barnard's arm. He repositions his hand low behind the table, so that only Barnard can see it. Fist clenched and thumb forward, he motions toward the screen. Barnard looks from one to the other of the projected couple, as he presumes is Frank's intent.

"I'm not sure if I have. All in all, I'd prefer to stay here," Frank says to the figures on the screen. "I need to find out about it. Maybe go for a visit, a real visit, not a 'net-site tour." He pauses. "There's something odd with the Elder Edens," Frank concludes. "So I'm not sure. Probably won't, to tell the truth."

Karl and Yelena's images seem to freeze. The on-screen pause that follows Frank's lengthy comment is like the inconsequential but noticeable stasis that can result from a momentary loss of downlink signal.

"Well, that is fine, Frank," Karl says firmly, becoming again a fluid visage. "Yes, you are correct. I am sure you will see it to be a good idea. We do. It is certainly for the best."

Yelena duplicates Karl's gestures, seemingly seconding his suggestions. It is as if they are trying to convey approval, their firm conviction, in fact. Barnard eyes Frank, simultaneously comparing the mild incongruity of Karl's previously relaxed manner with his current formal tone.

Vertrottelt sind? Gone dotty? No. Not that.
Has Frank noticed?

This final segment of the conversation is not as much puzzling to Barnard as it seems designed. He continues to study his friend, who is saying his goodbyes. He looks beyond, to the closed door of the VieGie room. Shadows can be seen on its panel of obscuring rain glass. He hears indistinct conversation.

"Kurchak visit complete," Frank declaims as the dissolving pair on the screen are in the process of getting up to leave. The screen blanks for an instant then presents the VieGie sign-in menu. Frank presses down on his knees and lurches up, catching Barnard's eye as he does so. With a slight movement of one eyebrow as they head out, he cuts off any comment. They must not disrupt the schedule, it would seem appropriate for him to be attempting to convey.

Frank is suggesting more, Barnard divines. He had needed no help with the session after all.

CHAPTER 7

BARNARD'S VIEGIE

Barnard sits close by the window. He looks out to gauge the wind by the disquiet of the palms then tilts his head downward to scan the street below. Doing without window screens makes the breeze and the view seem more accessible, unencumbered in a pleasant way. It is safe; he is safe. Few insects fly up as high as the fourth floor, he has heard people say. He cannot state this as fact, but he presumes it to be true since he has received few bites of contradiction throughout his years at Sunset Vistas. It was notably different on Second Street. There, even a tiny tear in a screen was sufficient to permit entry of faintly buzzing bitches.

Hands relaxed on his thighs, he sniffs up a hint of the ocean. It blends with the odor of subcontinent spice from one of the apartment buildings upwind. Barnard notices that it is nearly five twenty, seven twenty in San Angelo. He has scheduled his VieGie so that the grandkids will have spent time with their parents and not be impatient for their dinner. Kids always want something to eat, he recalls. In that there is certainty, which thought elicits a corresponding desire of his own. What, he wonders, might he snack on. Sunday lunch, downstairs, was thoroughly unappealing. Barnard had left most of it on his plate, as had Sallie but not May. He feels in need of something.

Cheese? Right, some cheese would be good.

184   7 - - BARNARD'S VIEGIE

"No crumbs," he murmurs, as he gets up and backs away from the window, maneuvering its blind to block the glare. He turns the straight-backed kitchen chair about before shifting the smaller of his two low tables closer to the big lounge chair. In the kitchen, he takes out a block of cheddar from the lowermost drawer of the fridge. He slices several thick, anticipatory slabs. No crackers, the retired economist decides, making the marginal concession in light of calories, salt, and fat, as well as later cleanup. A glass of water, to sip on as he visits, would be a good idea as well, he decides.

Lecturing for hours at a time. How did I...?
What was with Frank? Why me?

Whether conducted downstairs or up, VieGie visits are easy to initiate, are about as complicated as starting up a movie. Frank should not have required help with his this morning, but he had asked. Then he demonstrated that he had no need. Both were and still are puzzling. Frank's decline would be painful for Barnard, a significant loss since he has ties to so few others. Suppositions aside, Barnard prefers to think that Frank had something in mind, something other than a real or imagined need for assistance with his family session. Frank and his brother were talking at cross purposes by the end of the visit. He seemed to want Barnard to witness this but offered no explanation. Barnard senses a partial answer and regrets that he had not been more attentive throughout the entire session.

He peers at the red dot under his living room screen. In daylight and fully awake, he is not tempted to try to make it vanish by staring at it. But it does verify that the terminal is active although on stand-by and, he is inclined to acknowledge, that it is aware of him as well.

*   *   *   *   *

Few, save exhibitionists, welcome being continually observed. However, it is generally accepted now. Compensatory benefits include: noted availability for reminders; likewise for important notices, personal as well as public; the design of well-crafted opportunities, mercantile and otherwise; the occasional truly relevant announcement or alert. Furthermore, the interactive Internet and its remote visitation application, VieGie, are compiling an archive for the future, a resource for family and friends. All Web activity – be it for purchases, reservations, searches, events, appointments, enlightenments, or entertainments – has been tracked for decades. There is, thereby, ample motivation to cooperate with the steady accumulation of personal data. Everyone would have kept detailed personal diaries if it had been so easy.

Being googlified was Phil's clever summation. Likening the oversight to SORM, he had gone on, the ancient Russian System for Operative Investigative Activities, can make it seem ominous. "But it's not," he stated flatly during his dark themed harangue. "It's simply a matter of keeping track, of tabulating what's going on. Our lives consist of sequential experiences and memories are just linked data. The pokey-poke twits love having it set in stone. Stupid sots."

Barnard wishes he had not prompted his friend to start in on that particular tirade.

In any event, surveillance may, for some, imply malevolent intrusion but does not establish it. The activity can be supremely innocent. While Barnard understands that this interactive aspect of the Internet has a purpose, he is still not convinced that it has an aim, even after his revelatory experience of last year.

One evening after dinner, he, with some help from Sallie, moved his tall dresser to a temporary position in front of the bedroom Internet screen so they could shift his bed around. This was so they could try sleeping facing the window. While she had agreed that this might be a nice change, it was not. There was little to remark upon, other than roof stacks in moonlight or milky city shine on low clouds. As for stars, they could capture only a glimpse of a particularly bright one. Having the Internet screen off to one side, instead of straight ahead, would be inconvenient, as they could tell without even having to turn it on. It was a grossly bad idea and they decided to restore the original arrangement in the morning.

During their breakfast, before they could get on with the task, two gentlemen in utility clothes arrived at Barnard's apartment.

"We're here to check on your network," they said upon entering. "You appear to have a transmission problem."

Regular surveillance is not as universally denigrated as one might expect. The proffered advantages have been so diligently elaborated upon that it is an accepted activity, its host of benefits overwhelming the negatives. More basic and deftly utilized is the fact that, when repeated often enough, the awareness of being observed fatigues, as does vision, smell, or any special sense. One becomes inured to intrusion, just as, after on-screen repetition, one does to competitive cinematic violence, fright, and simulated sex. Most seem to have come to accept having their minutiae – from condition at birth to school records, from job and financial data to consumer interests and medical charts – retained for collation. Critically important transactions, for example those relevant to credit or to health care, do undoubtedly benefit from prompt access to accurate personal information. Less clearly of merit is the attention paid to the myriad of an individual's incidental behavior, commentaries, communications, and contacts. On the plus side, these accretions have made societal management easier and have given a welcomed boost to mercantilism. Unfortunately the surveillance of public and private spaces has not reduced crime, contrary to what had been projected. This failure is as much due to the ennui and aimlessness endemic among unskilled, virile young men as to misapplication of the diligently harvested data.

People adapt. Internet habitués have always been notably blasé about super cookies, for example, that made their data exhaust a valuable commodity for others. There was only transient dismay when it became common knowledge that many popular commercial programs – for text, photos, even financial tasks and security – incorporate surreptitious 'back doors' to take advantage of the rapid improvements in remote data storage. Users' soft and hard data – a log of each search undertaken and site visited, a compressed copy of every file created or modified, of every photo inserted, transformed, or transmitted, of every message in or out, of every financial transaction or transfer – are opportunistically uploaded into archives, there to be decoded for review according to utility, necessity, or even whim. Protective fire walls are easily breached when the need arises. Few seem concerned; the functionality gained outweighs the privacy lost.

Active monitoring is a firmly established aspect of the agreement under which one uses the Internet and is hardly a cause for paranoia. Of the associated, specific compensatory benefits, the two most significant for Barnard and his generation of peers are the familiar Biograph and the more recent VieGie.

Prior to the former, it had been burdensome to record faithfully the details of a family, its progress and occasional regression. Few took the time to formally catalog their lives, even on those rare occasions when a snatch of oral history might be ripe for sharing. Far fewer searched through genealogical databases, retrieved documents, or queried relatives so as to compile an organized narrative. Age brings with it the press of undeniable mortality. Seniors do wish to tell of themselves while they still can. Yet, "Christ, more 'When I was a boy,' crap?" the young invariably moan. When a desire for documentary permanence does arise, it is often too late. Memories dim precisely when the need is felt most acutely. Details succumb to the crude shorthand of familial anecdotes, of which few are ever transcribed. Yet, these incidentals are precisely what comprise the whole of a life. Only notables would write, or could have written for them, the true tales of their earthly presence. Until Biograph, that is.

A collection of photos lacking legends has tenuous utility when there is no one to identify the individuals, to explain their significance. Biographs are the opposite of this. Each is a dynamic entity, not a static post hoc compilation such as a scrap book or video. Once initiated, a Biograph is regularly updated. Data derived from incidental observational reports and every scrap of Internet activity are automatically scanned for those details deemed relevant and appropriate to add to the progressively enhanced dossier. It is as if everyone has been provided with a dedicated biographer, one who faithfully annotates the events of a life for the benefit of heirs and society alike. Additional significance is appended after the fact, usually in concert with data streams beyond the personal. The experiences, virtually the mentation of an individual, whether meant to be shared or not, thereby are integrated into an ever expanding whole.

As keeping a scrapbook or making periodic videos has been obviated by Biograph, VieGie has made distance irrelevant when contemplating getting with a relative or friend. A satisfactorily realistic visit needs only the Internet terminal, or, if a more enhanced experience is desired, one of the VieGie screening rooms that most retirement communities provide as a matter of course.

Biograph and the Internet's remote visitation adjunct, VieGie, have been most actively pressed upon those in or nearing their retirement years. For seniors like Barnard, there is considerable catching up to do. For the relatively young, with virtually all aspects of their personal and commercial activity mediated via the Internet, the incremental tabulations of their lives, their biographic personae, are contrastingly forward looking. They need no encouragement, need take no specific steps to see that this is done. It is being done. Their detailed dossiers are being opportunistically compiled from the inexorable accumulation of sequential presents, becoming dynamically updated, automatically refined repositories that obviate cooperation. A Biograph is far more reliable and accessible than a spinster aunt with fading memories.

While he remains suspicious of the systematic collaboration among the Internet, Biograph, and VieGie, their joint capabilities have caused Barnard to appreciate – ruefully, to be sure – the opportunity lost by neglecting to take advantage of Aunt Laura's experiences while she was alive. The generations to follow will not suffer the same. All aspects of their lives will be retained, in principle if not in truthful detail.

* * * * *

Barnard stares at the screen on the far wall, which is fine for entertainments but only adequate for visiting. Dwarfed by the one in the VieGie room downstairs, it cannot fully replicate its profound authenticity, hence, illusion of presence. Unfortunately, the formality of scheduling conflicts with spontaneity. In any event, the optimum times for a visit, namely weekends and early evenings, tend to get booked early. The necessity also to predefine an expected duration is another and, especially for Barnard, a tedious constraint. He is content to enjoy a VieGie in his apartment and give up visual enhancement for convenience.

The numeric time display below his large living room screen shows it is five thirty and fifteen. Then five thirty and sixteen, seventeen, eighteen, nineteen, twenty. Following and repeating the enumeration inwardly, Barnard does not want to be early for his Sunday visit with Kyle and family. He points the remote toward the screen to bring up the VieGie scheduler.

Twenty-five, twenty-six, twenty-seven.
Less than a minute late.

"Cordner. Barnard," he commands as he leans back into the chair. "For Cordner, Kyle, San Angelo, Texas."

Thirty-eight, thirty-nine.

His daughter-in-law, son Kyle, and their two children come to life before him, the latter on either side of their parents. While not as dramatic as when in the big VieGie room downstairs, their virtual reality on Barnard's living room wall screen is sufficient to be engaging.

"Just do it like we rehearsed it," the pretty teenager is saying to her younger brother, who obviously does not welcome her advice. "You're hopping up and down, trying to be funny. It's not just for the little kids, Brant."

"Yeah," her sibling deigns to reply, obviously unconvinced. He stares out through the screen in front of him at his grandfather, past the common boundary between them.

"Sorry. I'm late," interjects Barnard, smiling. He appends a professorial cough.

Kyle looks out from between the bickering pair and patiently returns his father's greeting. The bulge in the pocket of his stylishly wrinkled linen dress shirt hints of potential interruption. Barnard remembers, as Kyle never will, how pleasant it was to be deliberately unavailable.

"Not a problem, Dad. It's okay." Kyle looks briefly to either side. "They're arguing over the play next Sunday, the play for the Pre-Teen Summer Class," he explains.

Deborah, the elder, is a confident fourteen, with long black hair that is too Jewish. Her recently preferred nick name is DeeDee. Just as well. Deborah was a name never to Barnard's liking, despite the faint honoring of his late wife. He had casually suggested Derika but to no avail.

Brant, twelve, will soon catch up with his sister in height and worldliness. Barnard had asked – lobbied, in fact – to have them add the d, to make it Brandt. His grandfather would have been sympathetic to the attempt, even while no doubt displeased by the issue. His daughter-in-law had refused. Barnard felt he had the right to some say, which, of course, he had not. Again, he at least got an appropriate initial.

Kyle's wife stares out of the screen and gestures a polite greeting. Barnard studies her, the downstream image with the firm flesh of real life digitally conveyed. She appears pleasant still – heavier, possibly. Video tends to flatten the curves of the face. It adds virtual weight despite the convincing illusion of depth.

She's always so quiet. With Kyle, the same?

Those people will never forgive, it seems to Barnard. But Kyle seems happy. That is what matters, so far.

"You're trying to be funny, and it's not supposed to be funny," DeeDee is advising.

"Yes, it is," retorts Brant.

"Yeah? Not the way you're doing it. Fun, Brant!" DeeDee proclaims. "It's supposed to be fun, not fun-ee."

Barnard is mildly chagrined that his focus is wavering so soon.

"You both look great," he interjects, attempting to reconnect.

Deborah assumes an adult pose. "It's not for making the little kids laugh, Brant. They'll laugh at anything," she says. "It's the older ones that should get it. We're doing it for them, mostly. And they won't get it if you keep making your lines sound silly."

She tilts her head in a manner that Barnard finds poignantly familiar. He explores her face, lingering for a moment on her faintly lifted upper lip.

"Okay, guys. Drop it," Kyle directs sternly. "You can argue later. Pay attention."

"Hi, Grandpa," they say in an asynchronous chorus, suddenly children again. Brant waves his hand in his distinctive, young lad style. Barnard relaxes. His eyes dart from one to the other as he tries to detect more than a trace of a Diane or a Barnard in either of them. He trusts that the parental honorifics were sincere.

"Hi, kids," he says. Then, after a short pause, he tries, "Glad to be done with school? What do you guys have planned for the summer?"

These, he reasons, are as good topics as any with which to shift their focus. Their squabble could spoil an otherwise pleasant visit.

"We're rehearsing the play," clarifies DeeDee, overriding Barnard's intention.

Her throaty voice belies her age. While only two years older than Brant, she projects significantly superior maturity and is thus the quintessential big sister. Brant, recently having embarked upon his own journey through adolescence, will appreciate that one day.

DeeDee. A D at least. But with hair like her mother.
What was their ...?

Barnard, with difficulty, is trying to visualize Kyle's in-laws, or at least recollect their names. The lapse annoys him.

-berg? -baum?
Later.

"Grandpa. Are you listening?"

"Yes, I'm listening, Brant," Barnard replies, realizing that he must have missed something.

"I don't see why I have to act so serious, have to be so afraid of being sent to a museum," Brant restates stubbornly. "It should be great for Toby. An honor."

"They're doing a play from the Thomas the Tank Engine stories, Dad. 'Toby Gets Left Out'," Kyle explains.

DeeDee turns to present her father with a double shake of her head. "'Toby *Feels* Left Out'," she corrects.

Warm, familial memories engulf Barnard. Father had read the same stories to him. He can recall reading them to Kyle and Katie. She was the Brant then, he reflects. She was the one to take sharp exception to the socializing homilies of which Barnard was then barely aware. The surreptitious moralizing was part of the fun, as with Aesop or the Grimm Brothers. Barnard had not paid serious attention to the structure of those innocent tales as he read them out, neither to their potential hidden meanings nor their significance. The compact admonitions, guides, and reminders were the dramatic fodder of otherwise trivial entertainments. Now basically the same types of stories exist in multiple franchises, each with developmentally specific lessons to be firmly conveyed at the proper time and in the correct circumstance:

"Really useful engines are really busy engines."

"Dirty cars from dirty sidings."

Barnard remembers his father grimacing and nodding at that one.

The recalled "A happy engine is a useful engine" channels Barnard's grandfather, Adolph. As does "James was shocked! A steamie friendly with diesels?"

"Bother, bother," his father would say, simultaneously serious and distressed, then the embodiment of Sir Topham Hatt. "'Every wise engine knows that you cannot trust freight cars'," he would quote. Having such advice now issue from a thin tablet must be far less personal and, therefore, far more authoritative.

Populist political rhetoric from the fringe, especially during campaign seasons, often takes precisely this form, with the nouns suitably replaced, of course, according to perceived threat or vulnerable target. Many decades had to pass before Barnard would admit to some understanding of the true lessons being presented in those children's tales of which many seemed so fond. "The medium is the message," as McLuhan said. Attitudes are passed on much as are noses. Not via the genes, perhaps, but by example and, conceivably, by memes. That Barnard, roughly three score and ten years later, can still recall these bits of situational dialogue is poignant testimony to the power of early molding through the use of attractively colored rail stock, stiffly articulated wooden figures, and simple, short plots.

For children. For the kids.

"Pfffhhhhhh," Barnard emits as he rubs his rough chin. "Is it one I read to you when you still lived out here?" he asks the mildly squabbling pair.

"Yes," "No," they say over each other, to which DeeDee appends, "Maybe. But I remember Daddy reading them."

"I'm Toby," offers Brant, as if that was not already made clear.

"Yeah. And you're supposed to *be* an engine, not bumping back and forth like you're *being* an engine."

Barnard fails to catch the full nuance of the distinction, as apparently has Brant. DeeDee seems to have an instinct for performance and a well developed sense of authority. A perfect combination, it strikes Barnard, in their present milieu.

"That's a pretty good role, an important role," Barnard attempts soothingly. "A title role. I'll bet you picked that part."

"It's okay," Brant replies. "Only, why do I, I mean why does he have to be so afraid?"

"Because that's what's written *down* for you." DeeDee exhales loudly. She has been listening to his complaints for too long, that and her sliding tone convey unequivocally to Barnard. "It's not for you to decide. It's the story. Just act out what's written."

It is clear to Barnard that Brant is not enjoying his time on stage. It could be either the specific role assigned to him or the fact of having to be anyone other than himself that he is rejecting, the discomfort of fearing that the two – he and the character he has been asked to portray – may be conflated. Then he should not have to do it, Barnard concludes. Kyle and the grandchildren seem to take Barnard's distant look as an indication that he does not remember the story. He does, however. He remembers it well.

A museum is being opened on the Island of Sodor. It will be a place for people to go to see old things, things that are no longer Really Useful. Toby, the old tram engine, takes this personally, is afraid that he is going to be installed in that place. So he takes on extra jobs – from Thomas, James, Emily, and the others of their story world – so that he can prove he is a Really Useful Engine, one that does not belong in a museum. It all works out because it was a misunderstanding. The Fat Controller had no intention of sending Toby off.

Toby, or not Toby.

Hah.

"But why does Toby have to be all afraid?" Brant complains. "Why's it so bad to be in a museum?"

"Because, like I said, that's the way it's written." The girl shakes her head. She has cast aside any pretense of patience. She is a DeeDee becoming a Deborah. "Besides, those little kids are mostly going to notice the costumes and the sound effects. They're not going to get that it's, like, a fable," she says with a hint of arrogance. "That's for the older ones."

With a slow excursion of his head, Barnard takes in each of his grandchildren in turn, comfortable that they will not divine his thoughts. He molds his features so as to project that he does understand, that he does remember. It comes to him that being a witness to this banter is as good a way to visit as any. He leans back and smiles at Kyle.

"James and the others make him feel so bad. It's not fair. It makes me feel bad," his grandson is saying.

"Hey, Brant. Like, you're not the only one up there," Deborah scolds in a huff. "Do it the way it's written for you. You need to act out your part and let the rest of us act ours. It's not your play."

"Maybe that's a good start, Brant," Kyle attempts to soothe. "You can use those feelings to portray Toby even better. Okay? Enough of this already. Let it be and let's have a nice visit with Grandpa."

"Maybe going to a museum's a good thing when you're not really useful," a petulant voice insists.

"That's enough, Brant," Kyle commands. "Leave it." He faces out from the screen once again. "How are things on the coast, Dad?"

Barnard's flashbacks must fade before he can answer. He would not have DeeDee and Brant dwell on the stories, if it were his choice. He should have paid closer attention to Katie when he read out the stories to his own children. She was the prescient one, the one who refused to enjoy the childish tales, who took them as justification for rebellion not cooperation. Innate iconoclasm, reinforced and amplified by adolescence, could explain why she was so firmly estranged even before leaving home. Barnard had been either too slow or too busy to attend to any of it.

"Yeah, and what if he found out they were making the scrap yard ready for more work. You think he should be okay with thinking he was going to be sent there?" DeeDee says, as if to offer an alternative that she knows Brant will not appreciate.

"Enough, I said, you guys!" Kyle barks, before reengaging Barnard. "How's everything going? Feeling okay?"

"Yes. I'm doing, uh, I'm doing pretty well."

With this the visit slips into the commonplace, pleasant enough on both sides and incorporating a litany of trivia to span the three or four weeks since their last VieGie session. The disparate jumble of incidentals, today as on other days, is what keeps them prominent in each other's lives. They discuss Kyle's work, then Barnard's views on the changes around Ocean Park. Reflecting their different perspectives, they share their fondness for the beach – how they, as new Texans, do miss it and how

Barnard would if he had to leave. Getting to a beach from San Angelo is nearly impossible; it is at least an eight-hour drive, Kyle tells him. Barnard hears in detail about how the kids did at school, what next year will bring. Barnard answers a warm inquiry about Sallie, glad for the implication that Kyle, perhaps they, is/are happy for him. Kyle relates that his company still has a sizeable domestic operation, which is atypical for business software and media firms. Rumors notwithstanding, they probably will not decide to close it down and reestablish offshore, he tells his father. Barnard can tell that Kyle is aware of a large gap between wish and fact. There may be changes coming to their lives in Texas as well as to Barnard's in SoCal.

"It's strained here as well, Kyle. It seemed a good plan to buy into Sunset Vistas. Too bad events have overtaken it." Barnard hears himself say in a ponderous tone.

"A good plan. Yes.... Have you heard from Katie?" Kyle asks out of nowhere.

"Me? Me hear from her?" Barnard unnecessarily repeats, trying to relax his unintentional professorial stiffness. "I'm less likely than are you. Right? I suppose she's still with what's-his-name. Haven't seen or talked to her or had a text from her for nearly a year. Last Father's Day, I think it was," Barnard adds, as an affectionate, deliberately opportunistic goad.

He sees Kyle wince and immediately regrets having uttered the comment, which has imposed cool obligation on a warm interlude. Any expectation of a physical remembrance being on the way has evaporated; Father's Day will be honored by a belated text or a 'net-card. He has no further need to check his package box.

"I'm not sure if I could even get to her on VieGie," he comments, to sidestep his self-centered complaint.

"Try 'Kat Foeman'," Kyle suggests. "That's how she goes now."

"Have you talked to them?"

While Kyle slowly shakes his head to indicate he has not, Barnard is taken aback to realize that, through his own blithely automatic use of the terminal plural, he has acknowledged her union. After an ensuing pause, Barnard detects an urge to talk with the grandchildren, to try to reestablish firmer contacts. He experiences a pang of regret at being so out of touch with their day-to-day lives. Presumptions and memories are scant bases for a relationship.

"Thanks for the visit, Dad," Kyle suddenly states, however, rather formally, partly in response to his father's distant expression.

"Right. It's been too long. I didn't want to interrupt your busy days."

"You're looking good," Kyle offers in return.

"I'm fine, thanks." Barnard follows this with a short laugh. "But Sunset Vistas seems to be having its problems. Most of the retirement communities out here are running short, cutting back. Many'll have to close or restructure, I'm afraid."

"That won't happen," Kyle scoffs. "Might degrade, sure. That seems more likely the way it'll go."

His daughter-in-law leans forward. "We've seen that in the posts," she says, vocal at last. "You need not ... worry," she continues in an uneven cadence. "There are the Elder Edens. Have you decided what you'll do?"

Barnard makes no request for clarification of the somewhat disjointed comment. He lifts his shoulders, not feeling nearly as relaxed as he is attempting to project.

"Do you have a spare room?" he asks, making certain to be sufficiently jocular to project lack of intent as well as to mask his mild puzzlement.

"No, we don't," she replies too quickly, "and ... we are not sure what is going to happen with Kyle's work."

She pulls back, away from the screen, becoming smaller. Her stiff manner suggests to Barnard that she recognizes she did not have to say any of that. Barnard regrets having brought up in jest what he long ago had so thoroughly discarded.

Kyle picks up the thread.

"Yeah," he says. "We aren't going to be coming to the coast this year. Have to stay close. They may need to shift a lot more of the work offshore."

"Does that seem likely?" Barnard asks.

"Hard to say. You have to make your best guess, keep your guard up, and watch for the sneaky take-down."

Barnard appreciates Kyle's reference. It reminds him of the advice he had long ago offered to a maturing son across the dinner table. He allows a shadow of that ancient image to flicker between heart beats. Kyle may also be cognizant of the symmetry. Barnard looks aside at the grandkids, makes a point of holding Brant's gaze during which he experiences a shiver of poignancy. One day Kyle will be similarly peering into the eyes of his own anticipated perpetuity.

"By the way," Kyle continues. "The Elder Eden people sent us a link to a video tour. I ... know why they sent it to us again, exactly. It is ... your decision."

Barnard is annoyed by the frequent gaps in transmission.

"Right. I'm looking into it," he responds.

"The last one we got – last week, was it?" Kyle turns to one side and gets a confirmatory nod from his wife. "The last one went over a lot of financial details. What would happen if you suddenly were to get sick or needed help. You know, if you weren't able to manage so well and such like that. If some, ummhh ... " His occasional interjections and hesitancies are much like his father's. "Anyway," he continues, "you're more familiar with all that than I am, than we are."

"I've gone over the info," Barnard starts. He prefers not to dive into the stark details just yet. He adds nothing further and hopes that his son will not take that path.

"We have seen the Elder Eden video. There is one out in Orange, near the university in Irvine, I understand," Kyle states carefully, in a somewhat stilted tone of voice. He looks aside at his wife to elicit her assent once again. "That might be nice for you. Have you thought to specifically request reserving a spot at that one?"

"Well, no, I haven't. And it's not by the campus, Kyle. It's in rural Orange County. But as long as you've brought it up, it, uh, it, uh ... I just don't understand how they can do it. The financials I mean. It doesn't seem possible that they can provide what they promise without going even deeper into an actuarial hole. Right? So I don't get the point."

The projected scene jitters. Barnard's attention centered squarely on his son, he discerns a slight hesitation.

"Well, it does appear to be a good arrangement," Kyle remarks, in that same voice. "Especially with the additional facilities and long-term care included. It is like you said a few weeks ago. Being away from the city makes it less expensive. So it has to provide big overall savings."

This seems a particularly odd observation for Kyle to make. Barnard is on the cusp of contradicting with, "I would never have said that," but holds back. In any event, the nature of the relaxed, family visit has changed. He is curious whether Kyle feels the same.

"That's the plan," Barnard says instead. "Last month, on our VieGie, you say?" he then asks.

"Possibly," Kyle replies, after yet another glance toward his wife. His manner of speaking is again relaxed and familiar. "You were going to an event, or had just had one."

"Sundae Nite, maybe," Barnard suggests. He reviews his recent past, trying to recall to which visit Kyle may be referring. "Right," he appends softly, certain that Kyle is misremembering.

"Well, we've taken the Elder Eden virtual tour ourselves. It's very nice. I bet you would enjoy having the beach close by. You should sign up and move there, relocate."

That pitchman tone again.

Barnard has no immediate reply. He has the feeling that he and Kyle have each heard what they should. Their familial smiles remain fixed.

"Sorry, Dad. We're going to have to cut this short," Kyle says when his hand darts toward his shirt pocket. Barnard can make out his stealthy manipulation to stifle the vibration of his phone. "This is a conference call from Bangalore that was delayed from this afternoon. We call it Big Bang," he laughs, to mask his slight discomfort with the distraction. "It's very early there, so I've got to get this."

"That's okay. It's great seeing all of you," replies Barnard. "We, uh, we can visit longer next time. Learn your lines, Brant," he says in mock seriousness. Then, to his granddaughter, "You're quite the young lady, DeeDee. Don't waste it."

Barnard is not sure why he added this last remark and wishes he had not. It is late and it is lame. His grin back at the group is steady as they say quick goodbyes.

The Yes/No reschedule tab blinks for the motionless Barnard. He terminates his constricted exhale with a pulsed hum through his nose, reflecting on Kyle and the grandkids. Relaxing his expression, he again tries to fathom if, indeed, he has forgotten something. He does not recall a VieGie when he might have offered his son – or anyone, for that matter – a favorable opinion on the economics of the Elder Edens. His head makes small, involuntary oscillations right and left. Comparison or advisement are each sufficient to ascertain that one has or has not forgotten. Forgetting is not itself the problem; it can relieve as much as engender anxiety. The problem is the uncertainty, the nagging hints of what may have been lost. Happily, the act of questioning or even merely sensing an incongruity is a sign of sentience. When that is gone, what is lost will not be perceived as such and the problem will vanish. His head steadies.

Nature has a way.

Barnard starts to get up to open the blinds, to let in the light and look out. But, he decides, it is quite pleasant to simply remain seated, perfectly fine to nibble cheese and sip his water. He cups his hand to catch any stray crumbs and scans the weather crawl at the bottom of the screen. Tomorrow promises to be a nice day. It will be a bright, clear day and he can expect a nice breeze. Yes, tomorrow will be a fine day to take a walk, to perhaps exercise on the beach, then sit for a while in the sun.

Tomorrow. Always a tomorrow.
Always?

CHAPTER 8

CONSIDERATIONS OF ELDER EDEN

The bus left him off at Main, a few blocks from the beach. It was an easy stroll through the haphazard park to the sand where Barnard is just finishing his solitary exercises. The well-rehearsed, stylized movements – his forms – have allowed him to be pleasantly self-absorbed. The rhythm of the surf, the quasi-periodic pulse of its concluding tumbles, was the extent of any intrusion.

He tenses to gain extra value from his final repetitions, the linked poses having been softened by the gradual onset of his seniority. To benefit from his periodic workouts, Barnard must be attentive to posture and exert the correct amount of opposing force. He feels more than hears the grinding groans in his knees, the crunchy snaps in his neck, the pops in his shoulders. His stance is not as secure as it once was. Because of the yielding sand he prefers to claim, not due to age. His exertions might appear smoother if he were on the concrete path behind. The downside is that then there would be passersby to attend to.

"Your range of movement seems normal," Barnard was told a year or so ago, at his last exam – physician-speak for "Good enough, for someone of your age." The assessment had provided neither relief from nor satisfactory explanation of the morning stiffness, the stab of pain in his lower back when retrieving a dropped sock, or the occasional throbbing burn in his sinister wrist when awkwardly twisted.

He is glad for this Monday morning's semi-solitude. Not often is it so. He feels less constrained by the incursions of age after his thirty or so minutes of mild workout. Done with revisiting his version of taekwando on the sand under a friendly sun, he stands still. He cants his head side to side to elicit another faint crunch. The breeze feels good. The pulsation beyond the berm is soothing and familiar, the gulls silent and swift, the breeze steady. Barnard welcomes all of it. The summer sun has climbed. The sand is noticeably warm under his bare feet as he scans the Strand for its denizens, then surveys the scattering of figures on the beach beyond. He turns his eyes slowly back to the right, over his shoulder to his thin jacket neatly folded on the bench, his shoes upside down on top in anticipation of any sudden gust. Across the street not far away, two figures in white dress shirts are surveying their surroundings, appear to be evaluating them. Barnard is not concerned. His wallet is safe in the pocket of his shiny sweats. He presses on the bulge, to be sure.

"Pfffhhhhhhh."

Barnard shifts his attention to the condition of the sea, its light chop. The tide is ebbing. The surf's dependable rumble and rush, carried by the breeze over the scalloped surface of the back beach, override the traffic, the occasional strolling declaimer, the happy laughs of children, and the chatter of their minders. Those busy finalities of the breaking waves are hidden from view on the far side of an unstable rise. If he walked to the sea and stood upon the berm, he then could watch the foam rolling up the flat, rich beige of moist sand. He does not. He has no need. Barnard has connected with the sea through its sound. He sniffs to engage an alternate sense, taking in a pungent whiff of seaweed left exposed by the most recent high tide. Soon birds will come to pick at it. Others will arrive to probe for the small creatures living out lives beneath the damp surface. Reality engulfs Barnard. He feels the breeze on closed eyes, tries to remember any specific year or day or hour. Nothing comes, other than a vague sense of having a past. Behind he hears the whirs of bicyclists passing in haste, the sharp crunch of their tires rolling over dry sand.

It is time to go. He tentatively crooks his left arm then lets it drop. He frequently has done this in recent months and it annoys him. Barnard had discarded – conquered, he would claim – the convention of wearing a wrist watch long ago, in respect of the technology of the hand-helds, the ubiquity of digital displays, and the rarely needed precision. He glances upward. The angle of the sun, a sailor's judgment of the proximity to solar noon, suffices for the moment.

Idiotic daylight savings. Why don't they ...

He sighs lightly once more.

When newly arrived at Sunset Vistas, he often would walk the entire way from there to the beachfront and use the bus only for the homeward leg. Except for those few blocks between Lincoln and Highland, the westward course was downhill and easy. He would patiently march up the short, initial slope then, from Sixth, continue steadily down, in typical beach city fashion, crossing Main Street several blocks further on, for the west-facing benches and the edge of the sea beyond. He attempts the full trek less frequently the past year or so, makes more frequent use of his bus pass. Then the obligatory, terminal march over the sand does not seem so arduous, does not drain him of energy. Time is not the significant issue. It is virtually the same whether by foot or by bus, unless he would happen to arrive at the bus stop at exactly the right moment. He can recall briskly walking the entire way, casting expectant looks back, pleased by not being overtaken. He could still make the attempt. Only, there are other factors in play. And limitations, if he considers the expedition fully. Today he caught the Ocean Park Boulevard bus at its nearby 11th Street stop.

Barnard senses he has had enough of a workout this morning. He takes a last look out to sea, at a rising swell preparing to crest, before heading back. The proximate cause of the subsequent crashing thump is unseen behind him. He bends down for a piece of dull white poking up through the fine sand. Dry particles fall away from the shallow curve of it. He tests the sharp edge of the broken shell, once part of a closed, hospitable pair. He drops the brittle residue and swipes his thumb across his fingers.

Huge creatures once.

Barefoot. Must watch where ...

He scans the sand as he proceeds to the edge of the narrow ribbon of concrete. Sharp shards like that could cut him if he were not careful. It has happened. When he reaches the elevated edge of the concrete pathway, he cautiously pauses before ambling across to join his jacket. He had been yelled at and had replied likewise to a fast receding cyclist a few weeks ago. How dearly he had wished for something hard to throw at the furiously peddling, convex miscreant. A resident, whose name Barnard has forgotten, had a shoulder broken by one. Afterward the reprobate – unfortunately a lawyer – demanded to be paid for his bent wheel! On this midmorning of a typical coastal Monday, the Strand is not in active use and relatively safe.

The bench with wooden slats faces the water. Its molded concrete sides are bolted into the pathway. The old movables, with their intricate metal work, disappeared long ago, either into closed off yards or taken for scrap by scavengers. Barnard sits, pushing his jacket and shoes to one side. Far ahead, the ocean seems to merge with the distant low clouds. He lazily links his hands behind his neck and leans his head back. Behind crimson closed lids, he is motionless for several minutes. It has been a nice outing, so far.

Lowering his arms, he pulls in his legs and looks to either side along the concrete Strand, which is sparsely lined with palms. His head pivots slowly, exploring their sequential drooping tops. He feels the sun explore his face. The sunglasses help; although taking his lightweight, brimmed cap would have been wise. Diane had always cautioned him not to get too much sun. He would oblige, but only when on the boat. He runs his hand lightly over his hair, which is still sufficient for partial protection, in his opinion.

Next time.

The breeze is steady, unusually brisk for this time of day. He can hear palm fronds rubbing. "The wind annoys the palms" he had long ago applied to the sound. Barnard still favors that pretension of poetry. It frequently comes to mind on windy days when he looks down on them from his apartment and hears the shudder. Here, at the edge of the sand, the bare trunks are outlined against a bright, pale aqua sky. Barnard's brow is lightly creased as he listens and looks about. Yes, there are children in the play area after all, but they are preschoolers and small-voiced even while frenetic. In addition, the breeze is carrying their sound away from him. He need not be concerned about that. There is something else.

Separation.
Have I forgotten something?

A thin figure is alone on a nearby, inland facing bench. From its contours and the neat, smooth hair, he deduces it is a girl. She has on jeans with worn knees. Her thin top further substantiates her youth. Both hands are busy in their respective SimuGloves. Barnard watches the encased fingers, their fluttering and poking in the blank space above her lap as she manipulates her virtual controller. She is evidently caught up in the imagery on her spectacle video. Its heavy frame seems old-fashioned. Except for the lateral protrusions, they could be a hand-me-down or secondhand acquisition, what a teenage Barnard might have affected in honor of a lost icon. Barnard's stare is uninhibited. The dense lenses are impenetrable. He cannot see her eyes but is confident that they will never engage his. He is perfectly safe.

The girl's quiet concentration does not seem so strange to Barnard. He has just concluded being similarly engrossed. Empathetic, he tries to deduce what game of luck or skill she may be attempting to master, who or what she may be trying to overcome. Whatever her task, it typifies the preoccupations of a generation far removed from Barnard's. He scans her from head to sandaled feet, lingering stepwise in casual curiosity. She giggles, her head jerks up then down, upon which her phalangeal sinuations cease. Her fingers are still as she angles her face toward the sky. An end, it seems. Barnard sees her lips move. He doubts that she sees the annoyed palms, hears the gurgle of surf or the slight sound of children, feels the breeze or smells it. She relaxes her head to gaze, he presumes, straight ahead. Then her fingers resume their circumscribed dance of choice and control. Barnard appreciates that there must be some aim, some point to her persistence. Or is the point precisely that there is no point?

A step above happy pills.

Hah!

Dynamic tension. What was I supposed to –

A bright laugh from behind the hedge bordering the parking area to his left interrupts his thought. Others join in. Barnard turns his eyes toward the four figures in bathing suits emerging from the green mass. They are close enough for Barnard to hear the scrape and slap of their loose sandals against the sandy walkway. They approach, searching seaward as if having arrived for some prearranged meeting. Each, save one, carries a towel rolled into a bundle, which no doubt encloses sunscreen, a hat, a pair of jeans, underwear, and perhaps a surreptitious bottle that the many signs forbid.

So what. Done the same. With Diane and ...

The tallest girl, her blonde hair streaked with pastel pink, laughs. She pulls her sunglasses from their perch in her hair, one hand saluting as she scans over the beach. She leaps up, pulling the open towel from her shoulder and making a partial spin as she throws it in the air. It falls leisurely onto her hands. Barnard is rewarded with a glimpse of a taut ligament high on the inner aspect of her thigh when she lands, slightly off vertical. She laughs with the others then whips the towel around her neck. They are young, sexy, and secure. Barnard steadies his gaze to appreciate their vigor. The lone boy – lanky, as tanned as the girls, and gripping his rolled up towel at the middle – lugs a large beach bag, which, by color and design, evidently is not his.

It is a revealing mobile tableau. Barnard averts his eyes as the young man appears to meet his glance. Or he may be looking beyond him, at the buildings behind, fixing their configuration in memory in case they decide to walk a long way up the beach, toward the pier. Barnard has done the same on his walks in the surf.

At the neatly spaced array of barrier posts the quartet pauses, standing in a rough rank parallel to the shore, then moves as one onto the pocked surface of blown sand. Their scratchy footsteps cease as they step off the concrete. The tall girl leans against a post to take off her sandals, linking one arm through that of a compliant neighbor. She takes only a few steps before putting them on again in laughing haste. Barnard cannot make out what she says. He can imagine. She is a tenderfoot. In a few hours they all will feel the heat of the sunbaked sand, even on covered feet. She grasps her boyfriend's wrist, pointing further up the beach with her free hand. Three sets of eyes turn in the indicated direction. Friends are nearby and waiting, possibly. Or they have spotted a favored locale on the broad extent of monochrome beach. The clutch of tiny figures far ahead may signify new connections to be made. Barnard will never know.

He enjoys a satisfactory, unobserved stare at the girls moving away. He would like to see them also on their way home, with fine sand sparkling amongst damp fair hairs on legs and arms. Only, he would then not have this fine view of their rhythmic motion over yielding sand. As if intuitively aware of his attention, one of the girls pulls the bottom edge of her snug suit out and over an indented, undulating contour. They are unconcerned, Barnard notes. He tries to catch a glimpse of the delectable, shallow dimples that should be above the waist of their skimpy beach attire. Putting his hand over the top edge of his sunglasses does not help. The sun is not at the right angle. He assays their profiles as the group wheels to their right: young, trim, agile. Three tight yet fluid bottoms. Three sets of shapely long limbs erupting from the sand. He follows their progress as they diminish. It is as if he is drifting backward, not that they are moving away.

A car? Privileged. Rare.

Without his noticing her get up, the gamer has gone. Barnard is alone, facing a virtually empty beach. A brief pulse of wind presents a faint mélange of old shellfish and seaweed. The combined scent is a common coastal annoyance that today is pleasant because it is so slight. It is relatively refreshing, he also notes, after the institutional odors of Sunset

Vistas' entry, of its dining and common rooms, of its elevator and hallways. It is nothing like the confirmation of personal space that greets Barnard when he enters his apartment. It is different, also, from the effluvia of the vehicles that engulfs him on his outings. He breathes in what there is, no longer able to recall the smell of his house on Second Street. Detecting a hollow feeling akin to hunger, he rolls his lower lip over his teeth and judges the shadows of the palms. He is thirsty as well. He looks skyward again, reaching back for something he cannot quite grasp or specify.

He scans the bright world in front of him. The breeze does feel good. Hands flat on his lap, he closes his eyes for a delicious blending of sea, wind, and sun before rising. There was a time when, from this exact bench, he could hear the diesel whine as a bus would accelerate away from its stop at Pacific and Main. He then would hurry and catch it as it paused before making its left onto Ocean Park Boulevard. Unfortunately, Barnard no longer quick marches so smartly nor hears so well. Even if his hearing were still young, the LNG buses are swift and quiet. It would be exceedingly lucky anticipation for him to catch one.

Well before noon. Ten thirty? Eleven? What was I supposed to –

The footsteps are brisk and catch Barnard off-guard. He snaps open his lids to see two young men hovering. Their dress shirts and slacks seem incongruous to him. Made tense by surprise and its companion, suspicion, he presses down with restless feet.

"Good morning," offers the taller of the pair, his voice conveying the smile otherwise obscured by the sun's glare.

Ties? Wearing ties?

They are too quiet and too close. Barnard cautiously shifts his eyes from the face of the greeter to the left arm held loosely across his belly, then to his opposite limb, with its hand thrust casually into a trouser pocket. In the next beat he takes in the expectant look of the shorter, rounder companion who, with a thumb hooked under the strap of a dark backpack slung over one shoulder, seems no less ambiguous. Had they been unkempt or even dressed casually, he definitely would have cause to be alarmed.

Should I?

"Isn't this a beautiful place to be?" offers the former.

Barnard makes a weak gesture of agreement. From their manner, as well as their dress, he supposes that they are visitors. He wishes the husky lad would speak so that some trace of accent might provide a clue as to from where. Barnard reorients his face to the breeze. His visitors smile steadily down on him.

"We've been here for a week and every day has seemed so pleasant. Is the weather always this nice?" As Barnard makes no reply, the smooth voice continues, "Yes, this is such a wonderful place to be.... Do you live nearby?"

Barnard reexamines them, trying to decide what attitude to adopt. He settles on terse conversational.

"Not far. That way, over the rise," he says with a jerk of one thumb in that direction. He wants them to grasp how pleasant it was to relax in the bright sun, to have been undisturbed. "Pfffhhh," Barnard emits and turns his head away to reinforce his preference.

"Yes, you are fortunate," the shorter lad echoes.

Barnard squints up at them against the glare. Is it he or what they intend for him that underlies their mutual, sideways indications of agreement? Barnard concludes that they are evaluating whether he is rich or poor, of the local gentry or an opportunistic wanderer. If this were Hollywood Boulevard or Melrose, he would have reached a prompt understanding of the reason for their approach.

What are they up to? Maybe I should ...

"Do you have a few moments? We'd like to hear what you think about ..." The tall youth pauses, noting Barnard's unreceptive gaze, then resumes nonetheless, "... about where you and our country are heading."

Barnard feels his tension melt. He should have realized: their neat dress shirts; their ties and slacks; the backpack; the smooth, clean shaven faces of a second decade; their relaxed, forward manner and polished black shoes. They want his opinion, not his property or his urges.

"No, I really don't. I was doing my, uh, my exercises. Tai chi. So I'm resting. You'd have better luck getting that from someone else," Barnard says. By reawakening the patient, didactic ease with which he had formerly answered basic questions regarding latent demand or surplus supply, he tries to overpower their superficial warmth with his own.

"Yes, it's very nice here," the soft-spoken young man states.

"Right. Usually is," Barnard offers back. He makes a dismissive motion with his head. "I'm just having a quiet sit before heading home. Enjoy your visit."

That should do it, he judges. Pleasantly direct. He stares out over the beach, locks his eyes on a distant coasting gull.

"We are here for at least another three weeks," the shorter, obviously less adept of the pair admits, almost promises.

They nod, in unintended synchrony. The secondary speaks with prolonged vowels, suggesting he comes from the urban south or the heartland, definitely not anywhere far to the east or west.

Say "football." Or "flat tire."

In a prolonged conversation Barnard might be able to deduce from which region has come this pleasant, almost cherubic young man. Barnard is hopeful that this will not occur. He focuses his attention on the welcome caress of the breeze.

"Yes," the persistent interlocutor starts, picking up the thread, "it's so beautiful by the ocean. We don't get to visit it often.... My name is John," he offers after the pause. "And he," he adds, pointing ever so deliberately at the torso of his partner, "is Andrew. Do you come here often?"

They wait for Barnard to offer up his name in return or, at least to reply. It is hard for him to withhold, to be so profoundly reticent. But it is equally hard for him, having been made aware of their agenda, to tolerate the interruption. He does not want to do anything to prolong it.

"As much as I can, right," Barnard finally replies. "To watch the waves. So I really don't, uh ..."

"Isn't it a shame that with all that's so good, we've been afflicted with so much that is bad?" The young man's voice is subtly firmer. "What is your name, may I ask?" He seems compelled to persist, trained to gain proximity by asking.

There it is. It's what they do.

By the beach in sunny Southern California. In ties and still pale.

Barnard scans the pair. "Bill," he decides to tell them.

The sturdier, less forward of the two looks to his companion then briefly addresses Barnard. "Hello, Bill," he says. "My name is Andrew."

New at it.

"Doesn't it trouble you, Bill, that our country has been led astray, taken from its proper path?"

John asks this in a double sense. He seems to desire to evaluate his target's degree of awareness and, at the same time, to establish a bond through use of the proffered name. He moves to one side so as to not have the sun directly behind. It is a deliberate shift, one designed to keep Barnard's eyes on him.

"I, uh, I try not to worry about that. I'm retired," Barnard states, before looking away as smoothly as he can manage and definitely annoyed with himself at responding so directly.

Make an end?

"Oh, we'd like to do that, Bill," the tall lad says with a hint of sincere envy. "To enjoy this beautiful place and not worry. Someday, I hope. But there is too much to be done. There's much that needs to be changed."

He takes his hand from his pocket and waves at his junior to set the backpack down, beside Barnard's jacket and shoes.

"Do you understand why this is happening? Why we're having this disarray in our country, the social and financial distress?"

"I believe I do. Right," Barnard says.

His smile into space now has validity and he allows it to expand. Having responded honestly, if reluctantly, Barnard could, indeed, tell them in detail, he muses:

It started with an excess of debt. For what purpose? they'd rightly ask. Financial gain, obviously. Debt recast as profitable paper entities, I'd additionally instruct them. As numerical ephemera totalized in computer databases as if they were physical, as if they were real, like soybeans or renminbi or oil futures, instead of fanciful creations of The Great and The Powerful. Wall Street alchemy transmuted debt from obligation to asset, I'd state, so that it could be deftly marketed. Financial firms merged and purchased and overpowered to form bigger and richer and more powerful institutions, each of which controlled more than served. The attentive pair would certainly listen. Then why did companies and banks and governments and currencies fail? they'd be forced to inquire. Leverage, I'd quickly reply, leverage and complicit complacency. They failed because of clever monotonic men, I'd endeavor to explain, who knew precisely how the process was destined to end and perhaps desired it to be so. Hearing this, they would learn. They'd be compelled to yield to my acumen, to recognize that their own didactics are futile, and to leave me the hell alone.

Barnard is conflicted by the opposition of preference versus vanity. The fact that he has, even briefly, seen this as an opportunity to practice his insights makes him wince. He is annoyed by his hubris and by having succumbed to the memory of prior position. As his visitors study him,

Barnard accepts the pointlessness of his considerations and just wishes them gone.

"... and so, you believe that you understand," the well-practiced one is saying. "You think that because our troubles are of the financial world that their origin must be there as well. You hope that when the fever of our economy is soothed, then we will be made well. These are illusions, Bill. Our illness is at the core. Fever isn't the sickness, Bill. It's a reflection of the evil inside. To treat the fever and let remain the poison from which it comes is to prolong the evil, Bill."

Barnard looks up, startled that the youthful proselytizer seems to have read his mind. John feeds upon this reaction, imagining that he has struck a responsive chord.

"You think you understand because it's what you've had told to you so often," he states with confidence. He has determined that he can get to the meat of his task sooner rather than later. Response – even if as rebuttal or merely physical – is easier to bridge than is passivity. It demonstrates a willingness to engage, or at least a moment of connection. Either way, it hints of easy prey.

"Have you never considered that our problems come from not having the proper type of leaders? That we need a government that will take us in a new direction? Lead us firmly in the proper direction? Lead us toward God? I'm sure this has occurred to you, Bill."

The breeze, which just a few moments ago was so fresh on Barnard's cheeks, is now harsh and irritating. He wishes both it and his visitors gone. Barnard looks along the hard path to the few seated figures far away, to distant solitary walkers. He wills his demeanor to testify that he has absolutely no interest in hearing of their well-crafted understanding. They then will recognize that he is not a candidate for their remonstrations. I am here to relax and enjoy; let me do that quietly and alone, he attempts to project.

What Barnard intends as rejection his visitors take as precisely the opposite. He is their common cause. Barnard begins to consider that he is mistaken to have expected that understanding, whether he does have it or can achieve it, has any real place in their ministry. These focused young men have little real interest in his opinions. Far less would they be inclined to hear their basis explained. They simply want to mold them, to own them.

"You must wonder, Bill, where this is going, where *you* are going."

Indeed he does, except not as they have been programmed to perceive. Barnard grins, this time with the lowered eyes of a cynic. Barnard appreciates that studiously polite resistance will never have the effect he desires. They are too well schooled in overcoming that. He focuses on the sand ahead, on its sparkle and the occasional agitation of detritus by the wind then looks to either side.

What else?
Walk away. Toward others. There. Or there.
Too far and too ... Engage them?

Barnard appreciates that to state his views would be foolish. Hearing but not perceiving, they would declaim in earnest, play out their practiced, facile pseudo-debate. He could play with them, ask them to justify point after point, go ever deeper into obscurity until a first cause, some absolute must be invoked. They would feast on this, the simulated insatiable curiosity of a child, the monotonic repetition of Why? that must sooner or later be cut off by Intuitive Certainty lest it be interminable. Barnard elects to remain mute. He stares up at them for a few seconds, then purposely looks away, intending this to be unambiguously dissuasive, to send yet another and this time hopefully sufficient signal that his plan for the day did not include being enlightened.

"It is like sin, you see," interjects Andrew.

Barnard's manner has prompted the young man to leap a huge gap in argument. The latter's eyes shift to John to show that, yes, he remembers being given instructions to listen only. He retreats a step; to concede that while unable to resist the urge to demonstrate what he has learned, he will hereafter refrain from overreaching, will not again abandon his role.

John, to Andrew's great relief, builds upon the latter's interruption.

"Sin is our mortal enemy, Bill. And there are those who can help," John promises, "those who can recognize sin, who can point it out and defeat it. What our country needs, Bill, is a government that accepts Jesus. We need leaders who understand that He has taken our sins upon Himself and died for us. Through this, Bill, only through this will our burdens of disaster, strife, and financial discord be lifted. We've sinned, so we are being punished. The answer, then, is simple, isn't it, Bill? We must reject sin, renounce it and all from which it has come. We need leaders who have put sin behind them, leaders who understand and accept that there's a better way, a best way to soothe our country's distress. You may not have said this aloud, Bill. Still, you can accept that I'm telling you what I honestly believe, can't you, Bill? That a best way is possible?"

The questions, as scripted, can only be answered in the affirmative.

Since Barnard offers no reply, they must decide if he is reachable. No one is beyond salvation; this is embedded in their doctrine and ingrained in their methodology. If he is senile, however, their theologic seed is being cast upon sterile ground.

"Yes. It can be frightening," John says, imparting his own interpretation upon Barnard's silence. "That's understandable. Everyone has had that fear. When you place yourself in good hands, in strong hands, then the fear is taken away. All you need do is learn the truth. Doesn't that seem possible to you, Bill? Doesn't it seem possible that there is truth in what I say to you? Doesn't it seem possible that there's a new Truth, that there is help waiting to be given?" He pauses, glancing at his partner, implicitly instructing him. "You seem like an intelligent, interesting man, Bill. What is it that you do?"

Barnard continues to stare out over the sand. He adjusts his shoulders, lets out a sigh, and settles back. Blinking slowly, he decides he would not be displeased by having them judge him to be impaired and thereby immune to their ministry. His head yaws slowly. He is alone. He is singular and unsupported. He fills his lungs, allowing his cheeks to billow with the prolonged exhale, and weighs the option of being physically demonstrative, of presenting to his unwelcome visitors the furrowed brow and tight eyes of confusion born of deficit.

John and Andrew exchange quick looks. They have asked a lot of Bill and he has not answered. Barnard dares to hope they are mutually concluding that, in fact, he is unable to grasp their clearly stated points. What he has not foreseen is that even if this is so, he will provide good practice. John retreats a half step and signs to Andrew to open the backpack. Barnard regrets that he cannot refrain from glancing in. A stack of pamphlets is no surprise, nor is the Bible, its black cover well worn, the gilt cross barely legible. Andrew takes out a clipboard, which has been stashed upright behind their materials. He flips over a few pages of the highlighted text, since he must choose the best approach for one who appears to be a reluctant but eventually susceptible prospect. Barnard clings firmly to his intention to give no hint of being curious about their message, the nature of the pamphlets, or what is written on the sheets of Andrew's clipboard. He can make out little, in any event, except that, centered at the top of each page, is prominently displayed the familiar New Nation diglyph. The paired N's cleverly emphasize a single descending diagonal, the top left of one pinned to the bottom right of the other. When the breeze flutters the sheets, he sees the unnecessary confirmation, the NNR in black cursive lettering underneath.

NuNats! New Nation Republicans.

The New Nation Republican party has taken on the trappings of religion and is making good use of Applied Theology, much as engineers practice Applied Physics. It took a long period of patient effort for their time to come, for one of them to be welcome on the bridge. No longer do crowds hoot their disapproval and yell, "No nuts, no New Nats" – not since the party secured a majority in Congress and put one of their own in charge. They are confidently gearing up for the next election cycle, anticipating further gains. In them, politics and religion are wedded, each in service to the other.

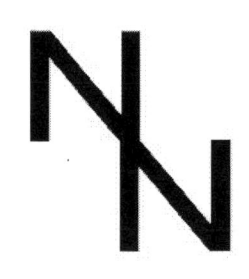

These two eager, loquacious youths are political, as well as religious, emissaries, two of the many trained to promote the good word of a revised political agenda and to spread the satisfying gospel of the government, by the government, for the government.

God "help" us.

Hah!

Barnard indulges his weak pun without any outward sign. It amuses him, but he also is on the edge of being angered. He knows that any smile or grin would be misinterpreted and only further encourage his visitors.

"Have you read much of the Bible, Bill? I mean, truly read it?" Andrew asks earnestly, coming forward. It is his turn to engage. "Do you know, for example, of the plan God has for us? For our country? Do you know that he sent Jesus, to suffer, to be crucified, and to be reborn for us and for our country? Do you know that His blood was shed for us, Bill? For us! For each and every one of us who believes! Is this not wonderful, Bill? His blood was shed to show that God understands, that He wants us to live in prosperous, peaceful times. Doesn't this prove that Jesus is here with us and wants ours to be the greatest country in the world? God sent Him and now has sent leaders to us, leaders who understand His promise and His goals for us." Andrew brightens as he receives the hoped for look of approval from his more senior companion. He holds up the black bound volume. "Let me –"

"Yes. The Bible. Right. It's a good book. Interesting reading," an exasperated Barnard spits out. He has sensed a favoring wind and an easy tack. Like a good sailor, he will put these to good use.

"History. It's history. Lots of knowing and begatting. Right? Right. Scads of miracles, rules, and predictions. There was a Moses, I'll give you that. And then there's that New Testament. There was a Jesus of Nazareth. I'll give you that as well. No doubt he was crucified; a lot before him were. Spartacus, for example. You've heard of him, right? Another political rebel, only not nearly so philosophical."

Heston? No. Fancy chin. Douglas. Right. The father, Kirk.

"Crucified, right. And a hell of a lot after. Criminals, martyrs, soldiers, captives, and just plain people. Poor, innocent, ordinary people. We've had many horrors come out of that truly ... good ... book."

This time Barnard's pauses are deliberate. He raises his hand to stymie any interruption.

"Whatever the Bible says, that's what you do. Is that it? You use it, mold it, pull out what you need to endure or to smite or to enslave. Right? Interpret, reinterpret, misinterpret – whatever it takes to justify whatever the hell you want to do. Whole nations have been put to the sword or pressed into being farm animals because somewhere in that fat book there's a passage that seems to say it's alright. *All* right. There's been a long chain of horrors from those days to this and we seem to be no better off for it. You have your plans, right. I'm sure. That's all we see in the media and on the 'net. Plans. And plans. And more goddamn plans. All to make the country sane again. Too bad they've been dreamt up by the same people who made us crazy. Right? You probably spend hours every damn day memorizing the illogic and made up stories then spewing it all out to strangers. Good luck. If you paid attention to the real world instead of letting your noses be pushed into your manuals and proscriptions then, uh, then ..."

Barnard stops, having finally run out of air. He enjoys having been wound up enough to deliver his unplanned diatribe without interruption, to have been able to run this far before the wind with it. The young pair's necks have stiffened. Andrew has taken a half-step backward, as if buffeted by a headwind, exactly as Barnard intended. The practice session has not gone as they wished. However, Barnard, who only wants to be left alone, is, at long last, going to get *his* wish. Passivity had not served. Logic was inappropriate. It was vitriol that worked, which is something for him to ponder.

John fixes Barnard with a cold stare then bends to fold over the flap of the backpack. "We'll be praying for you," he announces carefully as he picks up the backpack and passes it to his apprentice. He turns away and Andrew follows. Andrew pauses for one last look back at Barnard.

"Yes. We will pray for you," he echoes.

Their heads tilt in toward one another as they march north on the Strand, toward Santa Monica pier. Barnard lets his satisfaction show and craves that they should see it, while he considers which, from the long list of negative labels – generational, ethnic, medical, political, social – they are applying to him. He watches as they move away, tapping a light tattoo on his thigh as he seeks a fitting concluding phrase, those suitably final words.

"And don't forget for yourselves, while you're talking to Him. You'll need it more than I," the ruddy cheeked Barnard shouts to the departing figures. He wishes he had come up with that quicker. At least he has cast it upon them before they are out of earshot. He expects, inwardly commands them to look back. They do not. He urgently wants them to not escape his wry smile. But they do.

Black and white. Nothing in between. It's a uniform that suits them. Every damn one of them.

"Should be waving banners and parading in lockstep. Drums and bugles," he says aloud, with those small, involuntary head motions that attest to unresolved agitation.

Barnard finally relaxes. Only then is he mindful of the extent of his prior tension. He breaths and feels the warmth of the sun. Placing his hand over the jacket and shoes at the end of the bench, he pulls them toward him. He bends forward to slap away particles of sand from the cuffs of his trousers. Crossing one leg, then the other, he puts on his walking shoes, being careful to wipe each foot and to poke at the grit between his toes. He feels the throbbing in his neck. He still is breathing rapidly, as if having marched in quick time, although he has not taken a single step.

Shower. After lunch at OP. Pressbrot.
Walk partway there.
What was I supposed to ...?

After the pair are specks in the distance, Barnard again bends forward, this time to generate momentum for his rise from the bench. A biker speeds by. Erect, he follows the figure with his eyes then straightens his shoulders and strides forward.

Pains in the ass racers. Too fast. Why here? Should use the road ... Then THEY'd be at risk. Right?
Right.
Good workout. Not stiff today, after. Good.

There are joggers and walkers about. Few are attending to anything other than themselves and their gadgets. They seem to be unaware of the sun, the breeze, the sand, or the surf. It is a minor miracle that more are not knocked over by those on big wheels.

At Barnard Way he pauses before crossing, solely in respect of sensible caution. He no longer even glances at the coincidental street sign. That stretch was still Speedway when he was named. Traffic is light as he proceeds toward Main. He increases his pace, since the signal ahead is in his favor and a bus is waiting in the left turn lane ahead. Its stop is just beyond. He has a chance, he judges, if he hurries. He will take the short ride to the crest of the hill at Highland then walk the easy remainder of the way. That will further assuage his stress and be good for his appetite.

The ID-fier beeps acceptance of the pass safe in his wallet as he mounts the politely genuflecting, sandy steps. He chooses the bench seat across from the driver, facing the street, his hand instinctively reaching for the shiny post. He retracts it without completing the motion.

Who knows what ...

He scans the rows of worn, gray-blue seats and rubs both palms against his lap. Not many riders, he observes. It is neither the hour of going home nor of going to work. A couple – seniors but younger than he – nestle their large, grey cloth bags between their legs. A good time for grocery shopping, it seems. Coincidently, his one time housekeeper sits several rows further back. She is leaning close to the window on her left, perhaps idly studying the traffic or scanning signage on the long-closed branch library across the street. Barnard has no way to tell. Nor can he deduce why she happens to be on the bus at this time of day. She would normally be on a commuter's schedule: north, northwest in the morning, en route to her day at the Parks' condo on Nielson overlooking the beach; southeast, south in the evening, when heading home to her small apartment. After her misfortune in their hands, how she can remain in the Parks' employ, be of six days per week service to them, Barnard cannot begin to imagine. She does not see Barnard, which is fine with him. Or, she pretends not to.

Alece. Unusual.

The bus slowly picks up speed as it leaves the curb. An abrupt stop jerks Barnard to his right. He presses the side of his arm against the post to steady himself. The driver mutters something then, with a shake of his head, resumes his course toward Lincoln. Barnard rocks upright with the sudden acceleration, a barely conscious recovery, much like that during an over zealous breach. The bus swings past the offending (offended?) cyclist, whose middle finger salute is intended to reverse fault. A driver has to be wary, even the driver of a city bus. There is much fuss and paperwork if you hit one. Barnard has little sympathy for either.

* * * * *

It has been quite a few years since the Gold Park Bank almost collapsed. The Park family reached a negotiated settlement and came up with enough restitution to satisfy the regulators. That was after another of those public "show of guilt" trials, where the intent was to represent that agency staff were doing the job they were being paid to do as much as to harass the accused. Still, no amount of time or detailed complaints or minor acts of revenge could erase the bitterness if it had been he who had fallen into their inglorious morass of high-yield carbon equivalent tracking funds. Luckily, he was never tempted as was Alece. A hardworking woman with mixed-up English and anxious to provide for those years when she would not be employed, she had similarly aspiring friends who were even less discerning. None had imagined that the "funds" were Jung-a, aka Jules, Park's lucrative creations aimed at customers, his customers' friends, and whomever else he could entice without providing understandable disclosure. Yes, he got caught and was sentenced to time in a white-collar penitence camp. It was not enough. His incarceration was neither unpleasant nor of sufficient duration to tarnish his memory of the rich lifestyle that made it a fine bargain. Soon released upon appeal, he immediately took his place at the bank. Alece only marginally benefited from the delayed settlement, since most of the recoverables were absorbed by government fines and lawyers.

Alece has chosen to remain in their employ. Steady, relatively undemanding housekeeping work is too good a position to give up, even on principle – especially on principle, perhaps. Also, there could be that periodic extra cash payment, that surreptitious restitution in respect of cooperative silence, which, the former being unreported and latter unacknowledged, would bind her to them. Barnard suspected as much, from vague hints and incidental evidence. That Hutch Porter – Herbert is his birth name – had been burned on the same manner as Alece is harder for Barnard to understand.

* * * * *

8 - - CONSIDERATIONS OF ELDER EDEN

Barnard tries to suppress thoughts of Alece and Hutch, the one present before him, the other a linked image, as the bus continues its shallow climb east. He looks out past the driver. The traffic is light and there are few walkers this far from the beach. No one waits at the signed stops, no one on board buzzes to exit. The bus covers the several blocks in quiet haste.

"Pffhhhhhhhhhhh," he sighs.

Nobody. Good.

Goddamned NuNat Republicans. "Naries" a better moniker. Evocative but never stuck. Too bad.

Goddamned carbon chits.

Seeing that Highland is the next intersection, Barnard signals his intention by standing. His sunglasses, useful in the reflected glare from beach and sidewalk, impede any clear view of Alece as he steps down and out. Face straight ahead, his straining dexter eye confirms that she is still peering through the window. He expects that she will be getting off at Lincoln, to catch her southbound bus.

The bus speeds past Barnard as he proceeds across the intersection. The two long blocks ahead are downhill. He welcomes the pleasant, soon to be rewarded, mild exercise of a briskly paced walk. Ahead, the traffic on Lincoln's six lanes is not heavy. The signal already green in his favor, he determines that there will be insufficient time for a complete traverse by the time he arrives. He stops in the slender shade of a light standard, using the pause to look about and verify that Hutch is not at his post. He should not be, not until much later. Midday provides too few potential customers for the decorative paper favors he offers. The early evening, the hour or two of active homeward travel when the onset of leisure is foremost on the commuters' agendas, is understandably better.

The big man's image is unintentionally restored.

* * * * *

Herbert Porter, Hutch, is approaching his sixth decade and struggling to survive. As he is prone to complain, he is years away from getting any kind of government support, assuming there will be anything left when he qualifies. During the few years of occasional chats since making his acquaintance, Barnard learned that Hutch has a good grasp of financial matters and once had a decent income. Like many others, his intention to make dramatic gains worked against him. He erred, faltered, then lost virtually everything trying to recoup. His is only partially the story of aspiration, loss, and triumphant restoration. The last has eluded him and will forever be beyond his grasp. Barnard presumes that Hutch had put a portion aside. He is able to survive, after all. From conversations when his nephew Ty is not present, Barnard has learned that the latter's service medical benefits are a critical part of that survival.

With Ty's erratic help, Hutch forms sheets of bright tissue, twisted wire, and bits of colored tape into fist-sized creations intended as impulse items for homeward bound, busy swains, spouses, or friends. He seeks to tempt the commuters stopped at the light, to have them perceive his offerings as suitable tokens of friendship or as casually impulsive table gifts. As meager an enterprise as it is, selling the elaborate paper flowers does help Hutch's current finances at the margin. That is, the small incremental rewards are sadly significant. Barnard has often thought that if Hutch Porter would wear a sign board or put up a rough wooden stand, the comparison with images from the First Great Depression would be compelling.

As Hutch has explained to Barnard on several occasions, he has grown accustomed to his task. What initially was demeaning he has accepted as a challenge, one that has a number of secondary positive aspects: It keeps him and Ty occupied, and dulls any sense of total dependency. It allows him to feel that at least he is trying. Unfortunately, it also takes much standing and much patient acceptance of repeated rejection. It takes much mimed engagement and repetitious proffering toward closed or through occasionally open side windows. His efforts are a small step up from begging, but he does not regard them as shameful and is safe from being ticketed or arrested.

Hutch's appearance in class, long before Barnard retired, was immediately noted, because he was so attentive. Barnard used a nonconfrontational ploy to determine how he came to be there.

"Who's your advisor?" he asked. "You seem quite a bit older than the others."

"I've no advisor, Professor Cordner. I signed up to audit; no credit," Herbert Porter replied. He stated his full given name then added, "I do a little trading, stocks and commodities mostly. Want to get some background."

Perfectly reasonable, it seemed to Barnard, even if overstated. And, since no full time student was displaced, perfectly acceptable. SMC was a community campus, after all. He popped up in various classes, always attentive and always an active participant, apparently untroubled by his relative seniority. It was shortly after Diane died that Barnard recognized the vaguely familiar man at work across Lincoln from the OP, which was by then the saddened professor's frequent choice for early, solitary dinners. With a jumble of color in each hand and something like a flower poised precariously above one ear, he paced back and forth, gesturing toward the cars paused at the intersection. After a few opportunities to observe from a distance, Barnard treated him to a meal and learned the heavyset man's full story, one aspect of which was that his apartment on Navy Street was not far from Alece's pair of tiny rooms. That coincidence, that confluence of life's currents, enhanced Barnard's cordiality.

Living a short distance from Hutch and Ty, Barnard's former occasional housekeeper, Alece, would often see them at nearby Ozone Park, that cruelly named, narrow strip of tentative green where she would take her little dog to let it release energy after being cooped up for most of the day. She befriended the odd, lonely pair. On warm, pleasant evenings, the three would share a bench, Ty silent with feet outstretched, Hutch and Alece chatting. While she made use of the opportunity to improve her English, Ty would throw the ball, which he generously acquired for the little dog's amusement. He would repeatedly wipe it dry on his pant leg and soon tire them both. Paper plates of the Parks' leftovers, neatly bagged and stashed in Alece's big purse before she left for home in the evening, were occasional treats welcomed by the two men.

* * * * *

Yes, it is a coincidence seeing Alece on the bus just prior to arriving at Hutch's chosen corner. When the light changes, Barnard looks left to verify that the cars have heeded and notes that Alece is at the other side of Ocean Park Blvd. There she sits, on the sheltered bench of the stop. He smiles faintly. Facing away, toward the slowing southbound bus, she gets up to meet it.

Hutch and Alece. One intersection; two different lives. And now Barnard is traversing it as well.

Funny. Three of us. Odd funny.

He marches purposefully across the multiple lanes of Lincoln Boulevard. The walk sign is already flashing a warning before Barnard reaches the median. Lengthening his stride, he experiences a pleasant sense of release. He is no longer agitated from his encounter with the intrusive duo at the beachfront. Furthermore, he has been relieved of the tentatively anticipated obligation to exchange a few words with Alece. She was, after all, his weekly housekeeper on Second Street for several years.

Barnard enters the familiar OP Café. After all the talk yesterday, Sunday, coupled with today's unpleasant intrusion at the beach, he is looking forward to eating quietly. The available choices are at the utilitarian front, further in near the cold case and Max's station, or at the very back, adjacent to the voluble foursome who were loud enough to draw his attention as he entered. Barnard pauses. Like a chess player he evaluates the positional options: noisy trio, hectic work area, or constricted space by the cash register, front wall, and window where he is standing. One of the metal tables on the patio is not an option. He has had enough sun.

"Hello, Professor. They didn't wait to order."

Irene grins at the startle caused by her intrusion into his deliberations. Following her gaze, Barnard sees that Martin Stoole and Phil Winfree are at a table immediately behind him. The two servings of pressbrots that Max's waitress is carrying resurrect Phil's Sundae Nite suggestion.

Ahhhh. Eleven, he said. That's what it was.

Max's compact and satisfying pocket sandwiches emanate from an assemblage of open oven, link conveyor, and heavy steel rollers set up midway into the Café and well away from the window. Surely the front would be the better spot, Barnard had thought from the first. The process should be on display, should be a draw not hidden. People would be intrigued to watch the finale of Max Lohren's multi-step variation on pirogi. Barnard would not dare to suggest this, however. He has witnessed the man's temper. In addition, one is well advised to call it by its correct name. Max detests the Italian appellation, panino, and equally abhors any comparison with east European fare.

A complex assembly of ordinary ingredients, Max looks upon pressbrot as his unique creation. He loads paired, flat ovals of crusty dense bread with cheese and slices of the chosen cooked or preserved meat then adds tomato, sweet pickle, chopped pepperoncini, and pickled onion slices on one side, mayonnaise and obligatory, neatly trimmed pieces of red cabbage on the other. With prayerful hands he "presses" the assemblage together, then lets the metal belt convey it between radiant orange grids and pairs of slowly rolling, heated cylinders set barely an inch apart. Emerging from the apparatus super hot, the externally crisp concoction can be eaten either with utensils or as a tight sandwich. Barnard prefers the latter. If there are mysteries in the process, the primary must surely be why the ingredients do not squish out. Lohren no doubt spent much time and effort refining the spacing of the rollers, their speed and angles, the toaster temperature, and the timing. Barnard's curiosity has waned in recent years. He no longer feels compelled to watch.

Having turned about, he removes his sunglasses and looks down upon the seated pair.

"Well, hello. Sorry. Forgot what you said, Phil," he says. "I was at the beach, exercising," Barnard then explains as his gaze moves from one to the other.

Martin lifts his head and nods silently.

"Hey, Barnard. No big deal. Sit." Phil indicates the empty chair between them as Irene sets down their respective orders. "A day's getting too long for you to hold a thought?"

"Day and a half.... I'm going to have that," Barnard then says to Irene, pointing at one of the plates and otherwise ignoring Phil's friendly dig.

"Which meat?" she asks.

"Uhhh, pork. And coffee. Decaf. With milk."

"Max saw you come in. It won't take long," she says. "You gentlemen go ahead, before yours go cold," she adds with friendly familiarity.

It is not her place to offer that. Barnard, nevertheless, amplifies the advice with floppy hand motions as he takes a seat.

"And a big glass of water please, Irene," Barnard calls after her then glances at the plates on the table. "Right, Phil. I should have entered it into my packed calendar," he thereafter notes sarcastically. "What's going on?" he throws out as a perfectly good starting line.

"Not much," replies Phil. He shifts his plate closer then unfolds a napkin onto his lap. "We've been talking about the Elder Eden deal."

"Have you looked?" Martin wants to know. He picks up his tangy pocket of meat for a bite. Only then does he focus on Barnard.

"I have," Barnard feels pressed to reply. "Right. I, uh ..."

Lacking any conclusion, Barnard looks at the hand he has pressed flat on the table's edge then to the men on either side of him, of disparate ages but similar insularity. His eyes rest briefly on Martin Stoole.

No tan? Should have summer off.
Doesn't care, more likely.

"You've looked. At the site. So, have you seriously considered it, Barnard?" Martin asks as he chews.

Barnard did not and would not list Martin among his friends, barely as a colleague. He has always felt more comfortable keeping their conversations inconsequential. Consistent with Barnard's earlier estimation, he had turned out to be a satisfactory junior faculty member, one with adequate academic competence and adept at being never so categorical as to be subject to critical test. He has a soft voice, the kind he endeavors to make soothing and engaging. Unctuous is Barnard's less generous take on it. He was never warmed by Martin's presence nor by his efforts, which makes the current casual equality vaguely disconcerting.

"Well, actually I have." Barnard taps a slow rhythm on the table.

"It would be a good choice. At some point Vistas will have to drop most of its services."

Martin goes right to the core of their situation. He projects being less inclined to hear of Barnard's views then to tint them.

"When it does go under," Stoole continues, "it'll be an apartment complex. Just a place to reside. No meals. No services. You'll need to find a place that does have. Provide the housekeeping, the medical, and all that."

The resonance with the recent intrusion on the Strand perturbs Barnard.

A place to sleep, to eat, and to wait.

"*If* it goes under," he states sharply.

"Okay, Barnard. As you will. *If*," Martin scoffs back. "Still, have you ever pointed that out?"

"Point out? To whom?"

"Anybody. To the other residents, for example."

"I suppose. Not in any detail, really," Barnard answers. "Ownership's an issue. If they were to try to sell and go elsewhere, then they, uh –"

"What ownership?" Martin laughs. "It's a contingent annuity, not ownership."

Barnard looks straight ahead, declining to engage Martin's stare.

"Pffhhhhhhhhhhhh," he emits.

"True ownership's when –" Martin restarts.

"Guys. Can we change the topic?" Phil interrupts, noisily pulling his chair in closer.

Barnard feels that his former and forever academic seniority grants him the final word, that it outranks both Stoole's contrary insistence and Phil's apparent lack of interest.

"I understand that, Martin," he says. "I was going to say, that they've put up most, maybe all of their savings. And what's left is tied up for, uh, for what? Sixty years? A hundred, for Christ sake? Talk about frozen assets! It's a real problem."

Martin has stiffened and now dourly studies his plate.

Phil turns serially to each of his friends, making no attempt to mask his growing impatience.

"Let's talk about something else, or I'm going to go the hell outside," he says. "Looks like it was pretty breezy at the beach," he tries after evaluating his friend's hair. "I saw you come across Lincoln," he adds, to firm up the topical break.

Barnard has no need to reply.

The three make casual conversation – on the weather, current events, food at Lohren's, the kind of things that those lacking pressing tasks or responsibilities can indulge in. When Barnard's pressbrot arrives, he brackets the plate with his knife and fork, their handles equidistant from the table's edge. He carefully enjoys a hot first bite before describing a part, but not the major portion, of his morning. In light of Stoole's presence, he makes only casual reference to his recent offensive encounter. Martin, always quick to finish eating, does most of the talking. He sees fit to offer several short, illustrative homilies. Phil softens the thrust of Martin's observations with bland asides. Their conversation ebbing, they attend to the activity by the cash register.

"We don't, usually. I'll ask," Irene is saying to a customer.

She has no need. Max Lohren has stepped from his creative post to join them. He takes up the paper check the young woman has written, right in front of the "NO CHECK" sign.

"You write check? Ah. Zee zhere, vhat zine zay? You haf no account on your phun zhere? Miz ...?" He studies the check again. "Goldshtein?"

The woman makes several ambiguous gestures as she speaks, too softly for those at their table to hear.

"Vait. Vait. Okeh, you vizit. Zee here? You mek it Opies. Vhas's Opies? Zhis iz zhe Oh Pee Keffay."

Barnard tries but cannot see her face when Max tears up the check.

"Mek it tzu Oh Pee Keffay, den, or tzu Ozan Park Keffay. I neizzer care," Max says. "Opies? Nein. No. No Opies," Max says, leveling his gaze at the woman who is holding the check book cover open, pointing at it.

Last check. Check? Who still does?

Barnard and his companions feel free to stare at the shapely figure. The back of her light dress moves. It is a pleasant and thoroughly feminine back, Barnard observes.

"I tear up, becauze here iz Oh Pee Keffay, not Opiesss. I not care vhat you hear. Oh Pee Keffay not Opiesss!" Max hisses.

She gestures again, points outward and lifts two fingers.

"You vizit cloze. Okeh. You vizit. I neizzer care," Max says. "You haf no card zehr?"

She shakes her head and persists. Barnard can see the color rising on Max's thick neck. His pitch has sharpened a step or two.

"Zhen go, zhen. Go, not tzu come BEK!"

Phil and Barnard avert their eyes and exchange raised eyebrows, glad to be unobserved. Pure, rigid Max. Always the same. No one should dare label the focus of his life Opie's, as they should never be so impudent as to ask for a press-bread or for sliced turkey "on a French roll." Two of the three observers hunch forward over their secret grins.

Typical Max.

Martin rocks his head slowly. He does not share his companions' mild amusement.

"You need to be firm," he says, "do what you should. Even if it costs you."

Typical Stoole.

His is too serious an observation for the minor, comedic interlude. Yes, Max wants everyone to follow his rules. In this instance, however, it was not wise. Max did not gain anything by being so rigid. He lost. Tearing up the check was silly, vindictive, feckless. A proper businessman would have corrected the triviality and included it with the daily deposit. Max unnecessarily instantiated an unfavorable potential. Barnard shakes his head slowly.

Collapsed the wave equation.
Hah.

Martin had broken off his prior conversation with Phil when Barnard joined them. For his part, Barnard has little new for Phil in present company and nothing for Martin, whose presence is contrary to his preference. Therefore, soon after that one-sided drama has concluded, he brushes a few crumbs from his lips and drops the crumpled napkin on his plate.

"I need to go," he says abruptly, simultaneously signaling Irene. "Need to shower."

Barnard sweeps his check, which Irene has promptly provided, beneath the lens of his cellphone. Casually saluting the two men, he takes his leave. He decides to walk along Lincoln on his way back to Sunset Vistas. The car lot, the storefronts, the gas station, the traffic will be of marginally greater interest than the peeling garage doors and constricted side yards that border the sidewalk up along Ocean Park Boulevard. He comes to a halt at the corner and stares at the pedestrian signal as he waits to cross. When the icon changes to that stick figure with the scissor legs, he shifts his weight to his right foot.

"On your left!" he hears and jerks back. Barnard glares at the rapidly peddling, hunched-over man in the protective helmet.

"Stupid jerk-off!" he hears issue from himself.

He is dismayed by his tentative tone, by the lack of heft in his scolding. The biker raises one hand, then one finger, as he proceeds up the shallow slope. Barnard wishes he had projected louder, with authority.

"Where's a big fucking truck when you need one," he mouths. He looks up at the presumably watchful camera on the light standard.

Saw? Or broken?
Broken me, too, the prick. Doesn't care.
Max: "I neizzer care."

His head signs a resigned negative as Barnard restarts his quick march across Ocean Park Boulevard. He reaches the far curb as his allotted time expires and the blinking red palm steadies.

"Wish I had a stick," he mumbles.

Barnard is more watchful on the remainder of his walk to Sunset Vistas. Again, there is the sad line of slumped figures on the sidewalk in front of the clinic on the other side of Lincoln. He gives them a cursory glance, his thoughts disjointed, his habitual but haphazard count of his steps interrupted.

Angry, rigid Max. Grandpa Adolph, Aunt Laura, the same. Mother not. Not at all.

Max. Pure Max. Smart thing to do? No. No sense, no gain. Only loss. Cash the damned check. Anger misdirected. At whom? The unaware. Or the weak.

"I neizzer care." Hah!

No sense, no gain, just spite. Like the IOT vandals messing with door locks or freezing them. Making ACs run wild and burning out coffee pots, just because they can. Nasty graffiti. "Eat the rich" sprayed on a ticketed Maserati parked on Speedway. No reason, no point. Except for anger and spite. Right. And opportunity.

"Pffffffhhhhhh," Barnard emits in a long breath. He cannot help but shake his head.

Everywhere. Rules challenged biker bastards. Pushers and stinkers on line and on busses. Sneak thieves and scammers. And the NuNats. Right. NuNat Naries. Again. Right? Like generations ago, their chance's come again. Crazies? No. Not crazies. They're winners. Winners define. Victory confers the right, from which flows certainty.

Reaching Pine, Barnard turns right and marches up the slight grade toward Sunset Vistas. It is an easy walk for him today, he notices with some satisfaction. He cruises past the California bungalows with their unevenly kept front yards, their textured stucco stained from the damp. In a few years the single homes will be gone, replaced with boring pastel multi-units. That is what has happened on Second Street, he imagines. He had chosen specifically never to go by his old house. It might be gone.

Good workout. Tomorrow again?

Max. Should've taken the damn thing. No difference. "... neizzer care." Poor girl. Didn't know. How could she? A small sign, right. Hah. Entertainment for us, forty points up again for Max. Kill him one day. Maroon face, then bluish-gray and poof. Sirens and he's gone. "... neizzer care." Right. Poof and gone.

Hah.

Poor Father.

Fun seeing Max lose it.

Maybe we all should now and then. Might abort the trend.

Father thought he only had ...

Barnard's ruminations fade as he paces the last half-block to Sunset Vistas. Losing count of his steps, which is of no significance since he always restarts after "ninety-nine" anyway, he strokes the side of his head with an open hand then mirrors the smoothing motion with the other. Passing Phil's house, he takes off his light jacket, gripping it tightly, like a limp doll taken by the waist, and crosses the street between two parked cars. He wishes he lived closer to the surf. This far inland he only rarely can get a hint of it in the air. He has had few other regrets over his move to Sunset Vistas until recently. He had met Sallie, had privacy when he wanted, company when he did not. There had been special events at least once a week, escorted outings, special lectures. The medical staff was courteous and prompt. Even the meals, often a weak point at retirement communities, had made going to a restaurant an outing rather than an escape. It had been going as planned.

Right. Until recently.

Sundae Nites. Elder Eden.

Sunset Vistas' entry hallway and reception are relatively quiet. Standing in front of the elevator, he detects conversation behind him. He is insufficiently curious to listen. Instead, he peers at the numerals above the metal frame then steps aside to make way for a painfully thin woman exiting in a wheelchair. The soft-sided case on her lap jiggles as the rubber tires cross over the broad gap in the elevator's threshold. Tumph, go the guiders. Duumph, go the big wheels behind. Her wrists lie flat on her suitcase; her hands move restlessly as she is propelled onward. She reaches out with one hand toward Barnard, lightly touching his sleeve without effect. To Barnard it seems a familiar gesture. The attendant gives no sign as he presses forward. The two shapes accelerate down the hallway toward the ramp at the rear entrance of Sunset Vistas.

Tumph, duumph. Tumph, duumph.

He has recognized neither the resident nor the attendant. Someone from Zeven, he infers, as he mentally replays the indicator's declining sequence. Being fully dressed, with a knitted jacket draped over her shoulders, means she must be going out. The packed bag in her lap means she must be going away.

Relocating? To Elder Eden? Very likely. Right, right.

"Four," he commands, appending a professorial sniff of conclusion.

Name?

He watches them disappear down the slatted ramp as the elevator door closes. Tuuumph, duuuuumph, he hears faintly repeated in the distance. He puts out his hand to reverse the door's closing, thereby providing himself a moment to consider going to the common room instead. It is still early. He might find Sallie there, with May, no doubt. Without having made a firm decision, he lowers his hand and lets the door slide closed.

Safer to announce than touch. You never know.

Name? Her name? -stein? -stern?

His hands take on an unhealthy, bloodless shade, a corpse's color, in the cyan light. He holds them out, palms down in front, then lets them drop, limp, with fingers lightly curled. He rubs his thumb against the side of his index finger then along the seam of his pants. The dust-covered web in the far corner of the elevator has also lost any trace of color in the unnatural light. It has been a while since Barnard has had occasion to explore the upper reaches of this familiar moving box. Barnard angles his head forward, searching for the origin of the faint buzzing from within its strands. The web's creator, its arachnid resident, must be hidden in the gray mass, waiting patiently for exhaustion or venom to overcome its victim. Either way, there is no need for it to move quickly. Escape is precluded. A fly or moth, some short-lived creature with fast beating wings, has made a fatal error.

That is an imprecise projection upon the hapless meal-to-be, as Barnard is well aware. It cannot err, for it does not think. It is struggling because it seeks release, not because it is doomed. Like the spider, it is a living machine, a biological intricacy. Unlike the spider, its fate is to be reduced to a sac of nutrients. Significant benefit will be derived from its capture, benefit for another who, thus rewarded for patient labors, will consequently provide for her issue. The fresh morsel will thus serve a higher purpose than if it had played out the remainder of its own cycle uninterrupted. It is simply a matter of perspective, of many versus one.

* * * * *

Spiders wrap their prey, bind them after capture or a fatal bite so that they can be consumed at leisure. Barnard's father told him so. He also spoke of their many legs and intricate, species specific webs. And of their many eyes, which, with the proper orientation of a light in the dark of night, would beckon like tiny beacons. They would creep up close, the bright ovals from flashlights pressed against their heads contracting as they converged on the creature clinging to a stalk or blade of grass. It was always smaller than the sharp glint had promised. Whether far or near, their reflective eyes created points of light like grounded stars, and for similar reasons, as Barnard would learn in college a dozen years later.

Father also told him of the way spiders use their prey. A female spider –"the one with the eggs" was the way he put it on one particular pre-bedtime exploration – firmly wraps up any creature foolish or unlucky enough to wander into its web's sticky mass. She makes of it the larder from which the soon to hatch spiderlings will be fed. Enclosed in fine silk, the victim will be the nourishing host to a cluster of new ones that must come, that are coming from a previous dark union of which Barnard was ignorant. Thomas would jiggle the flashlight so as to point. "There. That white, round eggy thing at the edge of the web. The baby spiders are in there, with the dead bug." As Barnard then had no conception of the process of birth, neither did he grasp that they were observing one of the many facets of death.

These memories of evenings alone with his father remain accessible because there are so few. Thomas's mortality was extinguished before Barnard could learn to dread the possibility. When, much later, he learned of the way new life starts, he wished the revelation could have come from him.

Classroom instruction often was at odds with what had been imparted on those evenings. For example, not all spiders feed their issue on carrion, he eventually learned. Some capture and kill afresh for their offspring. Some produce or regurgitate food for them. There are those that leave their species' survival totally dependent upon instinct, good fortune, and immense redundancy. Whatever the strategy employed, their continued presence over eons of time is sufficient evidence of its utility, perhaps even incontrovertible justification for it.

*   *   *   *   *

Barnard casually scans the other corners. Not unseeable, merely unseen. And patient. That is the way of spiders. There must be others, he concludes as he stares upward. And there must be creatures in the elevator to sustain them. All must persist. Barnard ponders the incremental good provided by the creature being held fast in enveloping, gray threads. Those engendered will benefit. The cycle must work, else all would have been gone long ago.

"Four," states the elevator's annunciator as the box slows. The door glides open and Barnard stands motionless, waiting.

Two conjoining, then there are many. How do they do it?

"One," he commands and the elevator begins its descent. With a final look up at the web, the faint sound of struggle within masked by the noise from the first floor corridor, Barnard exits. He will search again, another time.

The still air in the common room carries the faint odors of flowers and old clothes. Even here at Sunset Vistas, many of the ladies will not go anywhere without a daub of scent. Barnard sees Sallie with May, amongst a cluster of unoccupied chairs at the far corner. As usual they are in earnest conversation. Sallie is of Northern Europe descent – slender and fair. She is fortunate to never have had to contend with maternal expansion. Her forehead is prominent and accentuated by her taut dark hair. Below are wide-set dark eyes and a narrow chin. May is of the South and notably heavyset. What had been a pretty, petite, round face – according to Sallie – now seems overripe.

Barnard was originally attracted to Sallie because of the physical suggestion of one intelligent and exotic, all of which is now overshadowed by how well they get along, by how pleasant it is to enjoy each other's company. That is the bond between them. He appreciates the unspoken yet substantially conveyed impression that she feels the same. As if to confirm this, Sallie brightens when she spots him. She signals him forward with a tilt of her head toward an empty chair. He is happy to accept the invitation, even though the wooden armchairs do not appear particularly comfortable. Each having found the other has made the recent few years so pleasant. And, by having May, who needs no other, Sallie requires no more of his time than he is willing to give. There is no need to voluntarily change that. Time will do it soon enough.

May, in mid-sentence when he was noticed by Sallie, continues on. Barnard sees her pudgy hands flutter across her ample lap, grasping and releasing each other. Rarely placid, her whole body takes part in any conversation, especially lately, with so much going on to rightly cause agitation. Barnard likes May. Still, he would prefer that she were elsewhere; Sallie's preference is distinctly the opposite. He is not perturbed. For each day there is a night.

Condi is by a window on the opposite side of the room, alone as always. She is smiling. Facing into the glare, Barnard cannot be sure at what or to whom. As a test he widens his mouth and gestures in her direction. She makes a small wave back. This, despite his intent, confuses Barnard. For an instant he toys with the idea of detouring over for a visit. He feels the flush of vague recollections, of what was and might have been.

Absence does ... Abstinence also. No, not today.

He will visit with her at the next Sundae Nite instead, he chooses.

Will it still be?

Barnard is close to drifting into another reverie, to conjuring up images of Diane, or of adventures before with those whose names, for the most part, are lost to time. He exhales through pursed lips and turns his attention to Sallie. May leans away from her as he moves closer and they both laugh. Their topic could be other than what Barnard suspects. He hopes so. He intends to avoid the serious, to decline any talk of Relocation. Flippant banter will serve instead. There will be plenty of occasions for serious discussion when Sallie and he are alone.

Following Sallie's gaze, May greets his arrival with a small smile. She forces her hands to be still, shifting position in anticipation of Barnard taking the adjacent empty seat.

Barnard raises his eyebrows ever so slightly, a questioning invitation unshared with May. Sallie's complementary nonverbal reply signifies "We're still talking."

"I got my letter of invitation, the official one, from Elder Eden," May announces abruptly, her inflection suggesting she cannot decide whether the missive comes as an honor or an imposition. "What do you think?"

"A letter?" Barnard asks with raised eyebrows.

"Well, you know, one of those long, personal net-mails. Is it a good idea?"

Barnard hesitates before replying.

"Well, we've all received those, or soon will." He pauses, again, this time a little longer. "I, uh, I suppose it's worth considering."

Barnard could hardly be less forthcoming. He does have much to say, but this is not the time any more than was Sundae Nite.

"It's nice here still. I don't want to relocate," May says as she trades glances with Sallie.

"Neither do I," Sallie affirms.

"I'd say the same," he says to them, as casually as he can. "There are people in Econ, at City, who are more in touch than I. I'll talk with a few, see what they think about the program. Right?"

"You already have an opinion. I can tell."

May has surmised correctly.

"I got another letter, as well," she then states, since Barnard does not seem inclined to offer anything further. In response to his look, she quickly adds, "No. An actual letter. On paper. In the box, under the robe."

"From her daughter," Sallie explains.

Barnard's gaze shifts from May to Sallie, then back again.

"Your children relocated to Chile years ago, didn't they?" he asks. "She's still here?"

"Only the boys. They seem to be doing okay. Yes, Roberta's still in San Diego. She wanted to stay here. There, I should say. In San Diego. She'll probably be leaving soon. Her rotten husband ran off. He had a good state job, as a computer programmer. I told her to get out before some little chippie, some *putana,* lured him away. Before he spent everything and maxed out their cards."

"Tell him about the letter, May," Sallie suggests, to keep her on point.

"It's a funny letter. Parts of it, at least. I was telling Sallie. She has funny stories about the people she works for. Roberta's gone to junior college and gets a salary. Still, they take her for a poor Mexican. Keep giving her clothes. Old clothes. Awful. 'Better than the Goodwill,' she says they keep telling her. And the clothes aren't even washed! They just –"

"May?" The intent of Sallie's insertion is obvious.

"All right. In the letter she tells me what she's been hearing about the Elder Edens." May frowns and turns more serious. "People she talks to, coming here through Tee Jay, hear a lot there that we don't."

"And?" Sallie again prompts.

Barnard also wishes she would get to the point. Drawn out detail blurs a story, any story, even a joke. Especially a joke. He folds his hands on his lap, crosses one leg over the other and re-engages May in mid-sentence.

"... is that their people at Elder Edens never get visited. They can't go see them. Why not? Why is that?" She hesitates, looking toward Barnard. "I know. They're far away and there's VieGie," she says, anticipating the comment he is poised to make.

"Right, May. They're on VieGie. So what? It's how they've set them up," he then tries to explain. "It's an advantage, actually. That way they can visit without having to sit in traffic or at an airport for hours."

"I know, I know. Set them up that way," she repeats. "But they can't have a regular visit. Not just like that. Not whenever they've a mind to. Can't even do a VieGie, 'just like that.' They have to schedule a visit exactly so," she amplifies, becoming increasingly animated. "Or, she says, they don't get through."

Barnard has heard this before. It does not sound unreasonable. Those who have relocated are fully available. A VieGie session should be prearranged because of the limited number of dedicated visitation rooms with life-sized screens. Access will improve as facilities develop, he is primed to predict.

"May," he starts instead, "it's no big deal. So they have a special room for the VieGie. It works better that way. We don't have huge screens like that in our apartments. When I want to talk to the grandkids, I usually use my own setup. The images are smaller so not as real as when I use the VieGie room, but it's no big deal. My screen's good enough for a quick visit. And I can do it on my own schedule. But if other people want –"

"I know. I know all that. It's that they don't *act* the same, Barnard. That's the other thing Roberta asked me. Whether I've heard that, that they've a funny way of talking sometimes. How they'll stop and stare out at you." May pauses, relaxes back again. "Anyway, that's what my daughter said in her letter. She has friends doing it from Chile even, and it's scary for them. They're afraid their parents have gotten sick and no one's telling them. Is that possible?"

"Anything's possible, May," Barnard says. He has heard others voice similar complaints. Realizing his is no answer, Barnard quickly adds, "The whole thing's so new. That's the biggest issue, I suppose. People get used to seeing each other in a certain way. When they see them another way, like through a video or holographic projection, it seems strange. You have to get used to the medium, the new way they're being shown. Right? Haven't you ever noticed that it works the other way around, too? Media people look different when you see them in person?"

While thin, his explanation is valid.

"There are technical constraints and development issues," he feels the need to append. Barnard nevertheless regrets this overreaching, since all along he has been reluctant to accept the mantle of authority held out for him. "It'll improve. The novelty of it is, uh, is the problem," he offers as compensation, immediately wishing he had not essentially repeated himself.

Professor Cordner does not believe a word of what he has said and is struggling to keep that from being apparent.

"Why did she write you a sneaky note like that? Put it in with the present?" Sallie asks her close friend.

Barnard, thus prompted, wonders the same.

"So, what do you think, then?" May says, neglecting to respond and turning in place toward Barnard. She appears to want a real answer, not serviceable public relations superficialities or misdirections.

"About VieGie?" he asks guardedly.

"No. About the Elder Edens. You must have an opinion. Should we go ahead and relocate? Things are getting terrible here."

Sallie, also, is awaiting Barnard's reply. May is correct. One need not be hypersensitive to notice the changes at Sunset Vistas. Many of the staff have left. Services have been reduced.

"I've scanned their Web site, that's all," Barnard at last admits. "The apartments seem nice. Nicer than here, possibly. I really don't have an answer, May," he eventually shrugs. "It comes down to a pretty much basic financial decision. Our money's tied up here. Soon we won't be getting much for it. So, if we can transfer those funds to a newer facility, one with some legs, as they say, I suppose it would make sense."

"Everything's a financial decision for you, Babe," Sallie observes, in an easy, accepting tone.

"Well, economic if not financial," he replies. "It's deciding how best to use what you have. It's the forever basic problem: Get the most for the least. It applies to most any decision. Right?"

Barnard feels that laudable compulsion to explain in detail what is overtaking them. He starts to imagine how he would lay it out. In chronological sequence, perhaps. That would be the easiest and make the most sense.

"We could stay here. Why not stay here?" Sallie asks. By her intonation she seems to be content with that.

May's look suggests she is not.

"I might leave and join the boys in Chile," she says.

"You're not going to get your money out of Sunset Vistas that way," offers Sallie, showing mild pique.

"Right, And not out of FTARP either," Barnard scoffs lightly, pushing his own musings aside. "What would you live on? Or would you move in with them?"

It is May's turn to be the focus of their attention. She relates how everyone said she was terribly unwise, years ago, when she insisted on putting their retirement money aside in bank accounts. Their friends and family, their employers, her husband even, protested she was being extraordinarily obtuse not to make use of tax efficient retirement funds.

"Then where did you put your retirement money?" Barnard asks, out of more than idle curiosity.

"Oh, I bought CDs for us. When they matured, I just renewed them. Kept on doing that. For a while I bought stocks and stock funds, too. If they did well, I paid the taxes. Then that all seemed scary to me. Lots of headache and paperwork. Lots of decisions. By the time I was widowed, the money was mainly just in the CDs. It didn't grow hardly at all; there was just the interest. But that was okay, really. It was safe. It was for when I got old that I had to worry. The boys were on their own. So was Roberta, actually. I used a good part of it for here, to buy my apartment. And because what was left wasn't in a retirement account, I still have what most people had taken over by that new law. Mmm, you know what I'm talking about. I did have to pay the IRS each year. That took away some."

Barnard finds it hard to reconcile her story with her plump, plain continence. She had ignored sound advice and ended up being better off than most.

"It wasn't tax advantaged, you mean," he puts in softly.

"What?" asks Sallie.

"With retirement accounts you either paid your taxes before you put the money in or when you took the money out. You didn't have to do both," Barnard summarizes, with his eyes fixed on May. "Not like with the Excess Assets Tax. But then the retirement accounts, uh, the tax advantaged accounts, were taken over by FTARP. It was easy enough for them to do; the funds were already segregated. EAT and FTARP aren't optional."

He looks from one to the other and decides to elaborate.

"Look. No one, at least no one outside of the Fed and the IRS, expected our retirement accounts to be converted into quasi-annuities. That's basically what FTARP, the Federal Tax Advantaged Retirement Program, does. It's not related to TARP, the Troubled Asset Relief Program of a couple of decades ago. Somebody's idea of a clever pun, I suppose. Anyway, FTARP's aim isn't sopping up bad assets. It's to put to good use a huge stock of stagnant assets." Barnard knows he is drifting dangerously far afield. "Right? Our contributory accounts, the IRA's and K's, are intact, we're told, merely restructured. It's just that all the funds not in a retirement annuity, like here, or not already in U.S. treasury paper, are now in sixty and one hundred year federal bonds. After all, the purpose of those accounts was to take care of our needs when we retired. Right? In principle this hasn't changed. FTARP simply was, is a way to legislate domestic demand for treasury paper – bonds – that reduced our dependence on foreign buyers. It's as if the Government itself bought long maturity bonds, except that under FTARP no new money had to be printed to do it! They ended up owning the underlying assets is all."

"Well, I suppose I understand all that, Babe," Sallie says. "I just hadn't heard that term, tax advantaged, before." She looks briefly off into the distance then toward May.

"Anyway," May continues, "I kept paying the taxes, on the interest. The money was never in a tax advantaged account, so it wasn't gobbled up by that eff-tarp business. Well, there's that other tax, Excess Assets, like you said. Still ..."

The irrational efficacy of May's actions shocks Barnard. How could that be? Had May allocated poorly and been lucky? Or was she a cautious, inadvertently farsighted soul in a thoroughly nondescript body? He puffs out his cheeks. Sallie grins, tickled to see that May has impressed him.

FTARP, right. Not wise planning. Still, in the end, smarter than I. Hah!

On many occasions – classes, Rotary and fraternal audiences, casual party conversations, with friends, and with his own son – Barnard had cautioned that management fees and taxes were the biggest burdens on conservative, long-term investment. It was important to avoid the former through judicious choice of low-burden fund families and to minimize the latter by maximal use of tax advantaged retirement programs. Where could there possibly have been error in that advice? What had he overlooked? Sovereign risk? Political mendacity? Foreign machinations? He had neglected to take into account any of these. He had let denial subvert suspicion. Barnard had not anticipated domestic application of the government's ever present prerogative to expropriate.

"Exigent circumstances" indeed. While an option for governments, a gambit for those having the power of the State, he never imagined that it could be personally applicable. Not here. Not in this country. Neither had he contemplated that tax and fiscal policies would become so incapable of managing the domestic budget that it would be necessary. He had not considered that foolishness, overreaching largesse, and political extremism would bring the U.S. government to the point of being dissatisfied with the incremental and therefore going for the whole.

Habitual neglect of or inadvertent blindness to unpalatable eventualities is what often makes ignorance yield a better result than does analysis. If one must choose between being lucky or smart, the smart pick lucky.

"Politicians. Ech. They snatched it all," May states flatly. "Grabbed it and wrapped it up for themselves. Like a spider."

"You're maligning the spiders, I'm afraid," Barnard says with a breathy laugh. "They're complex creatures. Highly evolved and quite beneficial. I, uh, I saw, uh ..."

"I still want to hear what you think," May persists, since it does not appear that Barnard is spontaneously going to offer more. "Is Relocation, going to an Elder Eden, a good idea or no? Is that going to be better than staying here?"

"I don't have enough information to say, May," Barnard replies warily. "The idea is, uh, is a good one, I suppose. Let's talk after I visit with a few people over at City College." He presses his palms down on his knees. "I'm going upstairs to freshen up and change. See you at dinner."

Sallie is well aware, because they have discussed it, that he has an incipient opinion regarding the Elder Edens, that he has multiple opinions on the entire sour situation into which they have been maneuvered. This is why she facilitates May's questions. However, he has chosen to deflect them. After he has had a chance to talk about it with Sallie, then he might be willing to share his thoughts honestly with May.

As he gets up to leave, he lifts one eyebrow and tilts his head, thereby conveying another invitation to Sallie. The tiny shake of her head is his answer. May and Sallie are evidently not finished with their talk.

"Maybe not today, maybe not tomorrow, but soon, and for the rest of ..."

Barnard is understandably disappointed. Nonetheless, he expects that they will have other opportunities to relax alone. Upon entering his apartment he does not react to Daedalus's greeting. He sinks slowly into the big chair facing the screen on his living room wall, only then gesturing the volume down.

The utility of the Elder Edens is a stated fact. Utility for whom? is the key question. He calls up the screen's menu and directs the cursor down to Go To, over to Previous, then Elder Eden. Barnard waves through the pictures. He expands one, skips rapidly past most, since he has seen them before on his small desk terminal. Even though set at a low volume, he can discern that the music accompanying the presentation varies with context. It is slow and smooth over the dining room images, rapid and bouncy when showing activities. Barnard enhances the audio when they show the patio and sunset views. There is that sound of the sea, much like what he might hope to hear through his presumptive window on a favorable breeze. It is exactly that hint of the surf that he loves.

How do they know?

The voice-over speaks of the Elder Eden's benefits, of which he has already heard much. Having no need to hear it again, he lowers the volume. He pages back to the introductory image and studies the arched and imposing entry gate, which stands at the head of the drive that leads up to the building. Again he sees the indistinct line of text in the metalwork of the gate – an identifier or phrase. Barnard points pinched fingers and flicks them open. He leans closer to the enlarged image. "From Belief Comes Bliss," its cursive welcoming, is then easy to make out.

Barnard fixates on the motto for a moment before exiting the Elder Eden site.

"Hello, Pro Fessor Cord Ner," intones the face hovering over his home page. "Please take a moment of Pause, to reflect with us and listen to the Word."

Barnard does not respond.

CHAPTER 9

CURRENT AFFAIRS

Phil pushes his handkerchief deep into his trouser pocket. "Do they always rush in that way?" he asks.

Barnard points his thumb back over his shoulder at the still active dining room. "The residents? Right. They do. Beats me why. The food's not that good anymore but still they come down early and queue up. Bored, I suspect. Too much time on their hands. Now there's a joke. Right? Too much time? It makes me feel old to see them waiting there with their blank looks, most of them."

"It's all relative," Phil replies with a shrug. "Anyway, we *are* old. That's a fact.... Thanks for dinner, by the way," he adds after a few paces.

The meeting room, this evening the venue for an hour or so of Martin Stoole's Current Affairs, is at the far end of the hall. Its doors are open. Barnard waves off his friend's appreciation.

"I'm glad you decided to come along. Sorry Sallie was so quiet," he says.

"Her friend May sure isn't," Phil observes.

"No," Barnard snorts amiably. "She seldom is. You can't blame her, though. She's upset over her children being in Chile and them not being able to visit. Her daughter has problems, too. A bad marriage."

"Yeah," Phil says. "Got that clear enough."

"She keeps asking me about Relocation," Barnard relates. "I don't know what the hell to tell her. You prefer your own little place, don't you?"

"Sure," Phil says. "It works for me. Besides, I've had enough of group life."

"It's not so bad, Phil. Not as good as it was. Right. But overall, it's not that bad," Barnard says, in a tone that suggests he is also trying to reassure himself. He adds nothing further. "You're head to head with Stoole a lot recently," he eventually comments. "I didn't think you were buddies."

"We aren't," Phil replies. "It's VieGie stuff. They've been asking me to check on things."

"They?"

"Yeah. Stoole. Other VieGie sites. Julie Chen, too, sometimes. I've been doing sys-main here, for Vistas, and places like this. It's extra money for me. Has other benefits, too."

"Well," Barnard replies after a pause, his curiosity partially satisfied, "that's not like what you used to do. Right? Doesn't working on VieGie require specific software experience, special training?"

"It does, sure, if you're a developer. I do maintenance, like I said. I don't write code, just do superficial, user type stuff. Test mods, glitches, local hardware checks. Integration. Stuff like that."

"You miss it, Phil?"

"Miss what?"

"Miss working. Your secret ops."

"Crap, Barnard. You kidding? I'd had enough. Besides, it was getting weird, as screwed up and twisted as Nam, and you know what shit that was. They were letting the DIE algorithm decide, to let it take the heat if there was a fuckup or blowback. Couldn't do that over there in the sixties. Had to hide it – any mistake – or make it look like some rogue dud's brain fart." Phil glances at Barnard and wrinkles his nose with a quick sniff, as if testing a pungent odor. "What I do with VieGie is pretty simple."

"And your project, tabletop sun burn? You don't talk much about that anymore."

"It's gotten sorta repetitive," Phil grins. "Most of the fun was in the preparation, the setup. After a while, after I ran out of heavy water, I ... It's mainly that I wasn't getting any data. Nothing useful at any rate. Tweaking this and trying that isn't as much fun as it was. What about you? You miss the college? You're still an academic at heart, that's obvious." Phil does not look to verify Barnard's reaction. "You miss being in a classroom and working with guys like Stoole?" Phil asks, with a forced laugh, to ease them off talk of his own deflected efforts.

"Stoole. Hmm." Barnard searches for a clever reply. "I miss him the same way I miss, uh ..."

"Having an inflamed butt?" Phil interjects.

"Right," Barnard says crisply. "You got it. That's it! A pain in the ass. Hah!"

A scattering of voices comes from the room ahead. Phil lifts his chin in that direction and asks, "Has he ever brought it up? The time you ripped him a new one in class? Has he ever tried to get back at you?"

Barnard studies Phil with friendly suspicion, not sure whether to laugh or not. It was a tale better enjoyed after a few inches of scotch.

"I sure would've," Phil goes on, in response to that look. "I wouldn't have let it pass."

"No. Never said a word about it. I never brought it up, either. Not directly, at any rate. You're right, though. I did go kind of overboard."

Phil sniffs firmly while reaching for his handkerchief. He does not get it out in time and sneezes into his palm.

"Sorry," he says. "Excuse me. Allergies."

"Gesundheit. You sure? Sounds more like a cold. It's easy to pick one up around here. Anyway, about Stoole, no. Never a word from him about it. I tried to forget it, too. Right? I was embarrassed, afterward. I'm glad we never had to talk about it."

"So, you strolled up to him and started yelling?" Phil says with a snort. "While he was in the middle of a friggin' sentence?"

He obviously relishes the image of a stolid Martin facing a bellowing Barnard in front of a classroom packed with the barely mature. The slowing pair face straight ahead as they approach this evening's gathering.

"Well, not that dramatic," Barnard replies. "He was turning back to the class, from the white board, as I came in. That was before we used tablets for everything. He didn't see me right off. The class did. He must've seen their eyes, because he looked around just as I got next to him. Tensed up like I was a DI." Barnard does at last chuckle at this and overlapping memories. His head oscillates slowly. "'Why has there been no midterm in this class?' I yelled. Really. I remember that I yelled. 'It should've been right there, on your schedule!' Blah. Blah. Blah."

Phil wipes his hand with the handkerchief. "Was it that important? A school requirement or whatever?"

"No. Well, right, *usually* the courses had midterms and finals. I thought he should've scheduled one, but he hadn't. Wasn't that important, really. It was his class. I guess I hadn't looked over his course outline before that morning. The truth is, I was already in a bad mood. Something happened at home. I don't remember what. I was walking along with those damn course schedules in my hand, looking at his, and it gigged me, I guess. I saw Stoole through the glass and barged in. Silly. One instant I was outside the classroom, then the next I was berating him in front of a bunch of kids. I can't remember even opening the damn door." Barnard emits a short puff through his nose.

"You must've."

"Right, Phil. I must have. Hah. Right. I smacked the schedules down on his lectern then kept slapping at them. I remember some of his notes fluttering to the floor." He shakes his head. "He must've been totally, uh ... Right. It was funny to me by the time I got back to my office. It wasn't to Stoole, I'll bet." Barnard's thin smile fades.

"I believe it," Phil inserts into the pause. "What could he do? It was his first year there. He had to take it, was probably afraid that he'd be gone if he didn't."

"No, no. Not his first year. But pretty early on, right. Before he got tenure. Diane didn't find it too funny. Neither did I after a while. It was embarrassing. Word got around. Buzz, buzz. Students are different now. They'd laugh at it or tweet it off to their BFs. They'd've whipped out their cellphones for a video then send it on and forget it. Hah."

Phil gestures that he understands.

"Pffhhhhhhhh," Barnard sighs before continuing. "It quite a while before I could talk to him easily. Oh, I'd call him in to go over unimportant crap. I had to try. I sat and had coffee with him a couple of times. Forced myself. He was never on my shortlist for conversation."

"That's kind of sad, Barnard. I'm surprised you were such a prick. Why didn't you just apologize?"

Barnard looks at his friend. "C'mon, Phil. I couldn't do that." Several steps further on, he adds, "It was silly. He just looked at me and blinked. Tried to keep his lecture notes together. Then later, when I saw him at faculty meetings or in the hallway, he, uh, he never said word one about it. Not one peep." Barnard pushes out another abbreviated nasal huff. "Stoole and his wife came to our Christmas parties, and you wouldn't have noticed a thing. Well, except for her. Always dressed like a pro. She was a real ... mmmmhh." Barnard does not finish, sucks an air kiss instead.

9 - - CURRENT AFFAIRS

Phil looks down and smiles, playing with the various potential implications of Barnard's final remark. Pausing at the doorway, they see some of the earlier arrivals still standing among the one-armed student chairs arranged in neat rows. Others are seated and adjusting the screens attached to their armrests or just staring at them. Martin Stoole is at the front of the room, limp hands dangling from thin arms.

"Are we too early?" asks Phil.

"No, we're not. It's seven thirty, isn't it?" Barnard looks toward the large, black, military style watch on the inside of Phil's wrist. "He never gets started right on the dot. It takes a while for them to settle so he can make his invocation."

They remain where they are.

"He does a pretty good job," Barnard feels the need to affirm.

"Yeah. He does," Phil allows.

"This is as much real exposure to current events as most get. It's too easy to get submerged in your own little world, to get trapped in Internet make-believe. Right? I follow some of the European feeds. There are still a couple that don't get edited down to pap." Barnard looks over at one of the chair screens, still blank. "Do you watch the 'net much, Phil?"

Phil nods a No. "I try not to get hooked on the damn thing," he offers. "When it lets me alone, that is; when I'm not checking out some VieGie issue. I read more than watch. Science stuff. I watch only when –"

Phil is interrupted by Barbarelli wheeling past. There is a rodent-like squeak of rubber wheels at the transition from tile to thin carpet. Tumph, duumph. They sidle over for him and watch without comment as he propels himself past the rows of chairs toward a clear space at the front of the room. Without waiting for Phil to resume, Barnard picks up his earlier dangling thread.

"The shows, the weekly stuff are a waste. So are movies, most of the time. I have my doubts about the rest of it, too. Even the stuff that's not *supposed* to be manufactured entertainment. Who the hell knows what's true on the news shows, the spews. They show what they want, not what is. 'All the Spews That Pass As News.' Right? It's gotten terrible since, uhh ..."

He stops, appearing to feel he has said enough. Phil neither agrees nor disagrees.

Barnard's cheeks puff out with another prolonged exhalation. He is still dwelling on that ancient memory, the Martin episode.

"That whole thing with Stoole burned me a hell of a lot more than it seemed to trouble him. And that," Barnard confesses, "annoyed me even more. Self-reinforcing. Right? People shouldn't be so easy going."

"Yes. That's feedback."

"What?"

"Feedback. Part of an output returned to sum with the input, so that the output gets even bigger." Phil looks aside at his friend. "I figured you learned about that at college."

"Nope. I was a physics major, not whatever-works-engineering."

It is an unnecessary, half-hearted dig.

"Yeah. There are different kinds of smarts. You had responsibility. You had, mmmh, I mean, you needed to do what ... Something else was bothering you, like you said.... Or maybe it just didn't bother *him*. Maybe he didn't answer back because he's not a sensitive guy," Phil finally manages to state.

Phil's focus has darted about too quickly for Barnard to make any cogent reply.

"Getting tenure was probably the uppermost thing on his mind," Barnard muses aloud instead. "If there were some rule, something written down, then probably he would've apologized, just to be safe. He knew there wasn't." Barnard looks down, reflecting back. "It probably was my lame excuse for kicking the cat. Right?... The funny part is that my blowing up like that, on a triviality, helped him get tenure more than anything *he* ever did."

The room is starting to quiet down. Barnard raises his eyes and lowers his voice.

"I felt guilty afterward. So I went way out of my way to tell the promotions committee how promising he was." Barnard chuckles softly. "'Promising,' my ass. He's no scholar. Never expected he would be. Not creative. It didn't much matter at SMC. Not then, at least. His courses went well and he got tenure. Barely. The students never complained. Some liked him a lot, surprisingly. Comical cartoons but no complaints."

"I learned pretty early on that the best tech's the one who never wanted to be an engineer," Phil says. "Happy to do whatever and go home. Maybe Stoole was being clever, devious clever," Phil dares to propose in addition. "Knew you better than you knew him. Saw what keeping quiet could get him. Ever think about that?"

"Pffffhhhhh," Barnard emits. He lets Phil have the last word.

The group this evening consists of still curious, relatively alert retirees, most of them residents of Sunset Vistas. Contrary to Barnard's dismissive view of Martin, his periodic offerings are popular. He presents understandable, albeit usually not much more than glib summaries of current issues and events. He collates and simplifies. Having an unimpressive curriculum vitae, a Ph.D. from a lackluster school, and possessing no real intellect, Martin was barely adequate for the brighter undergraduate students eager to transfer to one of the UC campuses. Except for Barnard's guilt imbued support, he would have been another of the multitude of six-and-out lecturers, another of the cheap-because-plentiful faculty who so densely populate higher education. He provides another fine illustration of how the course of an entire life can be defined by a seemingly momentary incident, a hasty or unwise decision – another's as easily as one's own.

Looking backward is an easy game of divination for late nights with alcohol. Deciding which of the many minor events of a given hour, a given day, or even a given era will later be judged decisive is another matter entirely. There stands Martin Stoole, firmly ensconced at Santa Monica College largely because of Barnard's irrational outburst and consequent perceived need for reparation. Professor Cordner had not anticipated longevity at the onset of Martin's appointment. Of the others whom Barnard recruited for SMC, there was that one who did seem particularly promising. Unfortunately, conclusive validation was not to come. Clifton Carter was dead before Barnard's prescience could be fully proofed, a victim of arbitrary incident who was outlived by the less deserving.

Barnard notices that, again, there are no cookies or cake. A few attendees are pouring themselves coffee from the well-used plastic carafe set on the table against the wall. All are waiting patiently. Thus far deprived of cogent explanations, those who have come share a need for answers, or better, for *an* answer, one singular and final. They are here to glean some sign that their prospects are not as dire as they have been told or beginning to fear.

"How can I ...?" Barnard overhears as they walk past the rows of media chairs. "... text they sent me said ..." he next picks out, then, "No. That's not what ..." comes from another quarter of the room.

The snatches have similar tonalities of concern. Even only partially heard, they create a coherent impression. Barnard shares their generation and, therefore, is sympathetic to their anxieties. Change is not easily borne, especially when options seem to have been purposefully narrowed. Nor is there adventure in change that equates to mere disruption.

Cutting back because one must is more stressful than when by personal preference. There can be a feeling of superiority, a perception of firm, even noble intent accompanying the latter. An entire population can take it upon itself to depart from the accustomed, therefore easily followed past. Their attitudes toward that which surrounds them can be altered from without, by formally uncommunicated yet common knowledge. The changes can be for ultimate good or bad. There often is no plan, no guiding principle. Rather, there is imitation, with the particles of like-minded decisions circulating amongst them and replicating like memes. This, in ways that are not always immediately evident, can cut both ways, like a two-edged sword. Therefore, since this informal intercourse is effective as well as sociologically inherent, it has had to be put under tight control, with protocols and authorities.

"Here," Phil says, when they are about halfway in. "In case we want to leave early."

"Typical undergraduate. Don't fall asleep as well," Barnard says with a brief laugh.

At the far end of the row is Doug Roach. Barnard stiffens and guides Phil to the row behind.

The chairs are broad-armed, metal-framed, and endowed with a thin rind of upholstery. A few of the network panels and track pads are set up to accommodate lefties. Barnard has never liked these media chairs. He sits, leaning forward for a moment, to preclude pressure on his lower spine. It is an unsustainable strategy and he soon must conform his posture to its support.

Everyone in attendance is disquieted by the easily discerned deterioration of their surroundings and retirement status. With what further changes may they soon be forced to contend? There are numerous examples – some ancient, some recent – of what happens when need exceeds means. In addition, there is the feeling, still largely unconscious, that they may not fully grasp what the underlying questions truly are. In addition to the known unknowns there are the unknown unknowns. The former can be managed; they yield equally well to either intellectual argument or to misdirection and soothing mendacity.

The latter engender fear, which is more malignant. As with the foreboding that immediately arises when hearing "I hate to tell you but ..." one tends to react before denominated specifics surface. Lacking concrete and shareable evidence, intuition – a maligned, subjective guide – is often relied upon. Once heavily discounted by those with a firmly objective orientation, it is now the target of intense manipulation. For those here this evening as well as everyone else, the young as well as the old, all of their interactions – with each other and with their families, with the institutions, entertainments, and services with which they engage – flow through a singular portal. All the advice and explanation, all the marketing, tempting, and coaxing, all the reportage and diversions, are contingent. If people are not prone to complain that what they receive is designed to match their specific, carefully cataloged lives, then most at least suspect it is so. The thematic interdigitation of that which comes downstream with what of them that has been passed forward, sent upstream to the central servers, is sensed even if not labeled.

The Age of Connection is a time of paradox. Instant communication via email and cellphones, and, most intensely, the compelling reality afforded by the Internet, its Daedalus Man, and VieGie, have bound those for whom the stated aim was to set free. Some future philosopher, one born well passed its onset, will have to dissect the powerful societal trends and provide their context in the past tense. Europe in the first half of the twentieth century required that sort of dispassionate explication. The financial manias, which periodically afflicted the greedy, the ambitious, and the innocent alike, deserved it. The jihads and pogroms and crusades and collapse of great civilizations demanded it. Trends must run their course before they can be analyzed critically and dispassionately. In contrast, for each of those sitting and waiting this evening, the Now necessitates individualized understanding not scholarly generalizations, practical adjustment not canonical truths.

The wrinkled and gray, Barnard's counterparts, those who have retained at least some measure of self-direction, are sensitive to disturbances of their personal space in ways that the young, and others similarly habituated to external manipulation, are incapable of perceiving. Youthful obstinacy is common, but the interior truth is that their rote rejection of authority is often a sham, a consequence of their acceptance of the authority of their own peers. It is not an honest, true act of rebellion. Compulsion and authority exist but are subliminal. These come via the agency of their constant commingling via social media. They delight in and derive a great measure of their security from reinforced identification with those of their generation. Yet, the young do not, cannot accept that they are managed ciphers. What they cling to, what they rely on is, must be, carefully orchestrated so that it seems otherwise. And thus it is, with the promptness that pervasive, instantaneous interconnection has only recently made feasible. The anticipation, the apparent plan, is that the younger generations shall follow the path laid out for them, that their underlying uniformity of thought will become the norm.

No societal action is without reaction, however. That is an imperative, much as in the physical world. What is planned for and implemented may be subverted in diverse ways, the positive as well as the negative. Either end of the moral spectrum could be in for significant surprise. Barnard has frequently sensed that much remains to unfold. Manipulative management, once the means are acquired, can have far more odious consequences than witless mercantilism.

He and his peers, in the straight-backed chairs this evening, are being offered Relocation, which is their primary concern. Via their presence at Current Affairs, they are demonstrating that they want to understand their options so as to make a rational decision. Putting into context Relocation's stated practical, if not financial benefits, requires more than what is provided on the Elder Eden net site. This fact is appreciated by those of the older generation. It is they who remember multiple reliable sources of news and opinion. The retired, the last of a once unconnected generation, remember how it was to argue over policy, to hear and to read competing points of view, to discern whorls of wisdom among a tidal churning of disparate ideas. This was part of the fabric of a variegated society comprising people who paid attention and were reasonably well informed, each of which attribute is being systematically diminished, if not already overcome.

Professor Cordner stares straight ahead. Martin's comments are anxiously anticipated by the group. His vague chin notwithstanding, he has an enabling tablet open on the lectern beside him. That he has a measure of utility cannot be dismissed. From Barnard's perspective, however, Stoole is biased and allied with the official doctrine. Therefore, his implicit agenda is suspect. One fact is clear: Who can say what conclusions those assembled might come to if they were not provided with the proper reality. To Barnard, Martin seems ill-suited to that role. Pale, thin, crudely mustached, his unpublished and certainly unacknowledged nickname comes to mind.

Professor Mutt. Hah!
Hairy mouth. How eat?
And not enough sun.

"Good evening," Martin at last offers to the group. "Shall we take a moment?" He then begins his invocation, speaking with great sincerity. "Lord God, bless this meeting and those gathered here."

Those present this evening are accustomed to Martin's spiritual formalities. To Barnard, they seem too designed, too precious to be taken seriously. Yet, as Martin directs his request to the assembly, as well as upwards, they attend to him, some even mouthing his words.

Barnard instead studies the screen on the arm of his chair, the gateway to the Internet. Links to helpful definitions, relevant facts, and illustrative examples are at the ready to clarify and propagate the agenda. The terminal is also unambiguously bidirectional. Queries, usually more authentic and revealing than the corresponding answers, will be systematically logged and collated. Via facile and flexible informatics, as is said of prayer, all will be heard.

"Our agenda tonight, as always, is God's. We seek His guidance, His help in understanding what is troubling us."

Barnard looks down at his shoes while Martin prays, trying to ignore the words but focusing on his inflection so that he might discern when they have reached the end of his beginning.

"I hope he doesn't drift into another talking head session," he confides to Phil. "Too much like a sermon."

"There's plenty of that," Phil agrees. "There absolutely is."

"Help us, Lord, to know and to accept Your wisdom. Help us to be good stewards of Your word and plan. We pray to You for our President and our country. We pray to You for those who would lead us to a brighter day."

He'd text it all caps.

"Most of their concerns are financial," Barnard leans to one side to say. "Changes in benefits. Changes in service. Their tied up funds. Eff-tarp is still a mystery to most of them. The rest is, uh ..."

"You could give them a good rundown."

"Right, right. Hmpff," Barnard sniffs. "I'm perfectly fine with not having been asked. There's too much to cover. When it's done, over, it'll be history. A full semester'll be about right."

With this half-hearted disclaimer, Barnard shrugs and turns both palms up as if to say: But, hey, look, here I am.

"... are difficulties that we cannot solve alone," he hears issuing from Stoole. "Therefore, we seek Your strength, Your guidance. We seek understanding and ask You to lead us to it."

9 -- CURRENT AFFAIRS

"Just do it," Phil says as he leans closer.

"Pfhhh," Barnard ejects. "No. No way. Not going to step on his tail again."

"It doesn't have to be like that. You could –"

"Shhhhh," insists someone behind them. They separate and succumb to outwardly respectful silence.

Martin concludes his overly long homily with, "We come to You, Lord Jesus, in our ignorance, and with humility, to ask for Your blessing on this discussion and the lives of us all. Thy will be done. Amen."

Barnard looks up at Martin and sees the supinated palms that seem to mimic his own gesture of a few seconds before. He winces at that similarity, as much as at Stoole's call for accepting and divining in place of thinking and observing. Critical thought is not apropos, it seems. He does battle with an urge to share this with Phil. Suddenly, Barnard profoundly regrets having stared down at his feet, is annoyed with his unintended pose of resigned submission.

Prays before every talk? God help them. Truly.

Barnard cannot suppress a throaty grunt. Phil mimes a request for clarification, which Barnard does not heed.

"You're aware," Martin proceeds, "of the financial dilemma that the Sunset Vistas retirement community, one of many, is in. The financial problems are not going to go away. Let's go over some of the related issues." Martin pokes at the keyboard. A terse, bulleted summary of his talk appears for each attendee.

The problems of which Stoole speaks have been building for several years and so have taken on an aura of familiarity. Martin, Barnard decides after skimming over the outline on his display, is going to play it safe. Facing an assembly of seniors, he is going to speak in general terms, provide a cursory overview and make use of the shared burden principle. There will be few attempts at specific answers, unless provoked, which Barnard doubts will occur. Barnard looks at the top of his screen, where "Elder Eden," in a curved line of large text, hangs over a diminutive view of the entry. The image is similar to that at which Barnard had stared at few nights before.

"Are you going to talk about the changes here? The lack of staff, the boring food, and all that?" someone in the front row interrupts to ask.

Barnard grins.

"Those are specific points. I'll get to them in a moment," Martin answers without direct engagement.

Barnard cannot see who has spoken. He guesses that it is Ted Bard, who is often quick to speak up. Having marked his territory, so to speak, he will pursue no follow-up. Martin is safe from again being pushed off-message from that quarter.

"I hope you've taken advantage of the Elder Eden site," Martin continues. Most have or soon will, wagging heads seem to suggest. "Has that been helpful?" he asks, more coaxing than questioning.

"Why's there all this talk about Elder Edens?"

It is not so much a question as it is a mildly frustrated rhetorical from somewhere behind Barnard. Martin obviously has heard, but his eyes do not shift in the intrusion's direction.

Hah. As quick as that.

"Like it or not, Sunset Vistas, or wherever you reside, is changing," he says, looking over his audience as if to establish how far renown of his presentations has spread. "It must. It's a financial matter that'll affect each of you in a variety of ways. Some immediate, some later." Martin scans his audience again, this time to gauge the effect of his assessment.

"Pfffhhhhhhhh," Barnard exhales. He makes use of Martin's pause to study the napes of those in front of him.

So damn melodramatic. As if they ...

Stoole turns slightly and taps the screen of his tablet.

"Sunset Vistas is close to being insolvent," he states. "Most of the other retirement communities are in the same predicament. It's a matter of finances, as I'm going to explain. The residences are in negative –" Martin stops and glances in Barnard's direction. "The retirement residences aren't covering their expenses," he rephrases. "They relied on investments that, unfortunately, have suffered severe losses and no longer generate enough income. They can't meet their obligations. The government can't offer much in the way of direct assistance because Social Security is already well into negative territory and is an outrageous burden on the federal budget. The growth of Medicare and Medicare Plus payments is unsupportable, as is the projected growth of Medicaid. There must be restrictions. There must be reductions in general and here, at Sunset Vistas, in particular. It's unpleasant, certainly. But it'd be far worse if the federal system itself were insolvent. As in a family budget, you can't spend what you don't have. Isn't that true?"

Barnard notes that Stoole's terminal observation is basically flawed. Governments, as is often the case with corporations, are not bound by the same financial constraints as are families, or most businesses, for that matter. On the other hand, to Stoole's credit, his summary of the current unpleasant state of affairs, while oversimplified, is apt.

"The cutbacks here are permanent," he goes on. "Services as they were are unsupportable, because funds are running out. It's not an easy problem to solve."

Or for you to explain, pompous twit.

"Sunset Vistas must either find additional income or reduce expenses further. To remain healthy, the government similarly has to control the entitlement expenses," Martin continues. "Relocation, the Elder Eden program, is a way to do both of these." He pauses, gambling that he will not be interrupted, and winning. "Elder Edens are being set up specifically to deal with the financial confusion and to provide alternative residential facilities. On the one hand, Relocation will reduce governmental outlays. In addition, it'll provide stressed retirement residences with an inflow of new funds through the arrival of new residents."

Barnard shifts in the too firm chair. He looks behind, to double check if Frank has deigned to come this evening. He would be quick to jump in with a Why now? or Why this?

"What happened to the Federal insurance programs, like the banks had? Isn't the money we put up here protected by, by, by some agency?"

Barnard cranes his neck to determine who has posed this. It is a deeply relevant question, one with a definite but not easily conveyed answer. He is curious as to how Stoole will address it, if at all.

"Well," Martin says, "that would apply to the mortgage here, yes. The money borrowed for improving Sunset Vistas was obtained through federally insured loans. There's a mechanism for handling that. The bigger problem is that there's more to supporting a retirement community than paying for the building. There must be operating funds, much of it local." Martin keys up a different image, a chart. "The money you invested in annuities here," he continues, "is no longer generating sufficient income to carry on as before. I'm afraid that means there's going to be a loss of some services, a loss of staff, maintenance, and extra programs."

Sundae Nites? What about Sundae Nites???

Martin pauses, inadvertently creating another opening for the attendees.

"Someone said that we're going to need to share apartments. Is that right? Is that what we −"

"It's not wise for us to get into rumors, Misses Farber," Martin says before she can finish. "No one is talking about that."

He has addressed one query by opening the gate to a host of others. Nicco Barbarelli, sounding primed and ready, takes advantage.

"We'v'eh heard that'eh we must'eh leave when we are'eh eigh'ty. Not'eh true?"

"That's also something to discuss. Only, again, it's too specific at this point of what I'm trying to go over," Martin quickly states, somewhat unwisely in Barnard's opinion. "Let me make my comments, then we'll go into that."

Barnard studies Martin's narrow frame, how it is accentuated by arms that remain unengaged as he speaks. He wears a dull sport coat over a patterned shirt with long sleeves that are not long enough. Tightly buttoned at the wrist, they transform his hands into ungainly extrusions. The lower half of his face is dominated by the dark mass of unkempt mustache, which moves with his words, not unlike a web agitated by a breeze. The remainder of his visage, from sharp nose past cheeks in need of a shave to the edge of his receding hairline, is pallid, faintly cadaverous. Overall, Barnard imagines that Martin Stoole resembles one of the grim, east European soldiers depicted in the undecipherable but temptingly illustrated "health" magazines he had explored in the storage area of his grandfather's house.

Stoole occasionally steps to the lectern to call up images for the chair displays. Barnard's grimace hardens as they serially appear. "Pace it out and space it out," he had advised junior faculty many times at summer retreats. Stoole's frames are too complicated. They illustrate, not his carefully choreographed key points, but the folly of overloading a single static visual, of doing what can be done instead of what should be done. They are useful aids only if one already knows what points the tables or bullet lists are there to make: Federal Budget; Treasury Market Investment Yield; Assigned Social Security Flows. Barnard can follow along. He has the unique advantage of having presented similar material many times. The current chart – Parallel Income and Expense Categories – is particularly overly detailed, with the respective shortfalls indicated by colored tops.

Barnard again squirms unobtrusively in his seat.

"Are they getting any of this, or only hearing it?" he whispers to Phil. "Expenses exceed income is no answer."

Specifically, he is curious why Stoole is slathering them with data. If he did more than tediously recite the obvious over a jumble of numbers, then they might grasp enough to be satisfied. To wit: Retirement centers are going broke; they were always dependent upon the government as well as upon their own financial base; both have weakened substantially; therefore, the funds that are available must be used more wisely. There it is. They need to hear of the nature of the shortfall and why the specifics of the Elder Eden program represent a rational plan for dealing with it. Instead, Barnard hears little that he imagines those here do not already feel, little that would engender cooperative understanding and a desire to adjust.

Martin drones on in the steady, patient manner of a preacher. Barnard, barely attentive, resurrects random headings from his own talks of decades ago:

"Economics is not science. It's the art of finding optimal choices and balancing competing needs."

"Expansion is generally good but requires innovation, new resources, or new credit, i.e., debt."

"Markets often go down faster than they go up."

"Investment and deficit spending each have different, forward-looking intentions, but each requires current outlays."

None of these would be helpful to the present audience. Barnard pries his attention away from the speaker and attends to his chair screen, where Stoole's current talking point is blocked out in neat bold type over a complex chart. Then he lowers his eyes to focus on the spot of light at the bottom. He attempts to stare fixedly so that it will disappear. The immobile red eye stares back. He cannot suppress the blinks and eye movements that deny him that insubstantial human victory. Barnard touches the keypad embedded in the broad arm of the chair.

"Welcome, Professor Cordner," in bold italics appears at the bottom of the screen, followed by scrolling smaller text that he does not read. He has been recognized and his attendance duly logged.

Barnard absorbs the current slide's title, "Elder Eden Offerings," and realizes that he has shifted to an overview of facilities. The images are different from those on the site he visited earlier, in his own apartment. There, what had been presented seemed real – posed, perhaps, but photographic. What he now sees are architectural renderings. Sketched, generic human figures impart the necessary scale. There are no views of the shore at sunset.

Could show a lot better visual than that.

"What you have had, here or elsewhere, was a very satisfactory arrangement," Professor Stoole is saying. "You received value for your money. Unfortunately the local residences are unable to deal with the prolonged contraction that we've had."

"And'eh the promise, Professor? What about'eh the promise?" Barbarelli inserts.

Martin declines to engage him directly. He evidently senses, as does Barnard, that it is less a question than a futile complaint issuing from evident fact.

"The promises, the agreements were based on things as they were," Stoole concedes to the group as a whole. "The situation has changed. That's the simple fact. We could discuss the whys for hours and it wouldn't help. What we need to talk about, and what you need to be considering, is what's being done to deal with these changes, what we can do together here to deal with the changes."

He pauses, pointedly avoiding eye contact with Barbarelli.

"There's no point to our going into the past, looking for reasons. You need to look at what is and not so concern yourselves with why it's different from what you had hoped for."

Martin flicks his eyes in Barnard's direction.

"Let's talk about the Elder Edens," he hurries on. "You've each received an invitational query. Many, most of you, have looked at the introductory video. They seem pleasant, wouldn't you say? They're stable and they provide significant advantages. Let me ask this: How does the idea of relocating to an Elder Eden strike you?"

There is a general shifting about throughout the room, a few tentative murmurings. To Barnard, Stoole's question is inopportune. Worse, it is misdirected. Barnard tries to decide if it is preachy admonition or glib sales pitch. Each is equally ill-advised. A sentence or two of explanatory background would be more useful at this point, something they could grasp and retain.

Stoole's manner, the entire setting, awakens Barnard's memory of the gratis, time-share "opportunity" weekend Diane and he thought would be a fun, cheap getaway. The pseudo-development, in Searchlight, Nevada, was an hour or so south of the glitz of Las Vegas. Apart from the bulk-mailed promotional fliers, the only way to be aware of its existence – the then Congressman Harry Reid notwithstanding – was the wind-worn, faded billboard seen when heading north to the gambling Mecca. Virtual captives for 24 hours, they had eaten bad food, played dollar blackjack at seedy tables, and listened to insistent, grandiose plans. "Never again," they had laughed on their drive home; "You get what you pay for." A slow warmth creeps over Barnard with the recollection. Nothing would ever happen in Searchlight, except for another, recently announced Elder Eden.

"Do we have'eh choice?" is asked.

Barnard, barely aware, unnecessarily leans to one side. It obviously is Barbarelli. The usually taciturn man is pressing forward on the arms of his mobile chair as he repeats the question with the same, flat intonation. Barnard emerges from his reverie to cynically consider whether Barbarelli is referring less to the Elder Edens than to the necessity of listening to the speaker drone on about them.

"Of course you do," Stoole replies. "No one's making the choice for you. They're providing the opportunity."

Several heads bob in apparent agreement. Stoole calls up views of common rooms then of a typical apartment. His flow is again interrupted.

"What about doctors?" Vera Hartmann attempts to ask. Barnard has seldom talked to the thin, nervously tart woman. She has straightened to cast her queries over the top of her screen. "Are you sure we'll have good medical care? That's very, very important. I've several medications that –"

"That's not an issue," Martin inserts. "Ample medical care is available. If a problem can't be handled on-site, then there's additional help on call. It'll be a big improvement." He pauses and looks from side to side. "You realize, I'm sure, that waiting times are longer and appointments are delayed the past year or so, that referrals are harder to obtain. Relocation will ease all that."

There is scattered muttering. Several related questions are asked in concert, becoming too intertwined for Barnard's imperfect attention to unravel. He searches the ceiling, sensing some truth in Stoole's assertions, and drifts to another place.

Martin is being afforded no such option. While he is, as planned, the focus of attention, his attempts to outline the general advantages of Relocation are being impeded. He would prefer to dwell on the wisdom of going to an Elder Eden over being pressed to address serial, directly specific fears. He raises his arm toward another questioner, trying to regain the initiative.

"That's correct," Martin replies, without restating what was asked. "The actual physical location isn't so important unless you, mmm, except when there are special needs or particular interests. For the most part, one's as good as another. The differences are, mmm, are primarily a matter of availability. That's the basic practical issue. That's what I believe should concern each of you at this point. Whether or not, I mean, the best apartments will still be available if you wait too long."

He seems to have been seeking the opportunity to insert exactly this point, to override anxiety with a sense of purposeful urgency. Martin keys up images of candidate apartments at the Orange County Elder Eden. He points out their features and makes note of their current availability. Barnard is barely attentive.

"... convenient also, yes," Martin is saying, in answer to another question. "Each Elder Eden has at least four large VieGie setups. This makes scheduling visits far easier than you are accustomed to here."

"Which've you visited?" someone from a few rows ahead of Barnard asks.

"None, of course," Stoole replies. "Elder Edens aren't set up for visitors during this startup stage."

"What about family? What if my sister or my nieces want to come?" a pleasant voice inquires.

Barnard attempts to associate a name with it. Whitten, Miss Whitten, finally comes to him. She is the "nice spinster lady" to whom he occasionally has taken a slice of cake or dish of ice cream after visiting with Condi on a Sundae Nite. Barnard is surprised when Martin addresses her by her given name.

"Well, Harriet," Stoole says, "they're located at some distance from here, by design, as a cost saving measure." He is obviously sympathetic to her question and to her as a person as well. This issue of visitation fits with his agenda. "It'd be a waste of time for relatives to keep going out to, for example, to the one in Antelope Valley. Elder Edens don't need to provide for visitors precisely because VieGie makes that sort of travel unnecessary."

"Elder Edens are pretty far outside the city. Okay. And we have VieGie. I understand all that," states someone ahead. To Barnard, it sounds like Al Silver. "But if they can put money into building a nice facility way the heck out there, then why don't they use the money to support what's already here?"

At last, the crux. Right to the damn core.

Lifting his chin to look through the jumble of interposed heads, Barnard notes that his identification is correct.

"We needn't go into that," Martin advises Silver sharply. "It's a governmental decision. I'm sure there are many good reasons. One is that a single large facility is more efficient than a lot of small ones. That's what's called economy of scale. It avoids duplication. You see?"

"Every three or four weeks I need to get shots. In my hip. Otherwise it's painful when I walk. Will they do that at Elder Eden?"

"As I've said, Elder Edens have on-site medical staff," Martin reprises. "All the medications or procedures you take or need here, you'll have there."

The attendees are drifting into personal particulars. Stoole holds up one open hand, trying to resist being nudged hopelessly off topic.

"When I bought my unit, I thought it was for as long as I wanted?" asserts a different voice, followed by murmurs of like-minded dissatisfaction.

"Well, that's still true. But we need to –"

"It'eh was'eh guarantee. For as long'eh as I need'eh," Barbarelli appends firmly. "Why'eh so different'eh now'eh?"

"As I said, that's still true. Exploring the history isn't going to help with deciding when and to which Elder Eden to transfer. An Elder Eden makes a lot of sense precisely because of the changes necessary here," Martin coolly replies. "That's what I'm trying to explain."

Barnard makes another lengthy exhale and leans slightly sideways, trying to take the pressure off one buttock. He is fighting an urge to grab Phil's arm and leave. He scratches lightly on the mouse pad to call up an earlier screen. There it is, the Elder Eden entry. He progressively expands the view, until he can easily read the "From Belief Comes Bliss" in the scroll work over the gate.

"It's'eh guaranteed'eh? Like'eh annuity?"

"Yes, it's guaranteed," Martin states. "And the funds are there to honor the guarantee. Moving into one would be a wise choice for you, for anyone eligible."

The interruptions, particularly Barbarelli's repetitious persistence, are taking their toll. Although Professor Stoole is endeavoring to appear unfazed, his slow walk to the opposite side of the room, his stiffer posture, his impatient eyes, suggest he would prefer to be done with them, with Barbarelli in particular.

"Government'eh, they guarantee?" The speech of the man in the wheelchair has taken on an increasingly evident ethnic cadence. "Like'eh they guarantee at'eh the bank'eh? They take'eh the I R'eh A account'eh and'eh then they say they'eh, they can no help'eh the Sunset'eh Vista? I don't'eh understand'eh. They have'eh money for this'eh and'eh for that'eh, and'eh not'eh for here?"

Martin is left with restating what, at least to him, is obvious.

"A promise is only as good as the ability to honor it. We have a crisis that must be dealt with." Martin looks over the group. "Sunset Vistas can no longer provide what it doesn't have the money to pay for," he says.

"I was'eh told'eh –"

"It doesn't make any difference what you were told, Mister Barbarelli. There are going to be changes because the financials here are no longer adequate. The funds to keep on as before are not, mmm ... The necessary funds aren't there."

"You mean'eh, I must to leave'eh here? Go to some'eh Elder Eden?"

"No. It's your choice. Only, it would be an obvious improvement over what you have here." Martin holds up both hands. "For example, let's look at the nearby Orange and Kern County Elder Edens."

"I do not'eh want'eh go to some'eh Elder Eden," Barbarelli insists. "Why do I need'eh leave and'eh to go there? What'eh changed'eh? What'eh went'eh wrong'eh?"

"Please. I understand. We'll simply have to go into that another time." Martin drifts forward, closer to the front row. He then redirects his gaze toward Barnard. "Even better," he continues, "possibly Professor Cordner could provide some clarification."

Barnard stiffens in his chair. He has been enjoying the relaxed amusement of Barbarelli's unabashed carping. Fortunately, he has remained sufficiently alert to grasp Martin's invitation, to not have to suffer its repetition. Its sincerity can be evaluated later.

"I worked with Doctor Cordner at Santa Monica College for a number of years. I believe he could explain our economic problems in detail, possibly better than me."

> *Worked with?*
> *Better than me?*
> *Than ME?*
> *POSSIBLY better than ...?*

"Pffhhhhh," Barnard emits.

"Would you be able to do that, say, at our Current Affairs on the ninth?" Stoole addresses directly to his former Chairman. "Perhaps you've already given some thought to this."

Barnard's impression is that the suggestion is a ploy, Stoole's attempt at misdirection or escape. Regardless, he quickly recognizes that he must respond. Besides, he almost welcomes the request. Having his fifty-five minutes in which to lay it out for them, in an understandable fashion, will complement his long standing project. He lets a few seconds tick by before indicating assent.

Good timing. For The Book, right.

"Yes, Doctor Stoole," Barnard confirms, aping the polite academic tone that Martin has adopted. "I'll be happy to do that. Two weeks from today? That'll be fine."

"Will that give you enough time?"

Prick.

"Oh, I believe so. A lecture hour or so should be adequate."

"I mean, will two weeks give you enough time to prepare?"

Sure you "meant," asshole prick.

"Yes, certainly, Martin. I'm quite familiar with the topic," Barnard says evenly. He trusts that his tone of voice is sufficiently close to friendly as well as professional.

"Thank you, Doctor Cordner." Martin says, with no hint of perceiving any diminution. "At the next Current Affairs, then, Professor Cordner will summarize the current situation and what that implies for the future," he clarifies, leaning deferentially toward Barnard with any indication of his true inner state hidden behind an ample, overhanging growth.

Barnard engages the several expectant faces that have turned in his direction. He feels that he can indeed make some sense of it for them.

"I don't'eh understand'eh," he hears from Barbarelli.

"Pfhhhhhhhh," Barnard sighs with a weak grin.

Tenacious old bugger. Won't let go.

"Profess'eh Stoole'eh. All'eh my children have'eh gone'eh, are'eh dead'eh. I am'eh alone'eh. All I had'eh was'eh my I R'eh A. This'eh the government'eh say is'eh gone'eh. Now, you say I must'eh –"

Martin repeats the gesture of holding up both hands, giving up any pretense of patience. "No. Please. As I said, this is not the time to –"

"Let'eh me ask'eh," Barbarelli interrupts in turn, his voice quite firm. "I come'eh here as'eh young'eh man. I become'eh citizen and'eh work all'eh my life. I put'eh'way money for when'eh I was'eh goin' to be old'eh. Now I *am*'eh old'eh, and'eh the government'eh take'eh what I've'eh, what I've'eh saved'eh and'eh –"

9 - - CURRENT AFFAIRS

"Mick. Meeko. We –"

"I am'eh NOT'eh Mick. I am'eh NOT'eh MICO. My NAME'eh is'eh Niccolo, Nee ko lo Bar bar elli." His agitated insistence makes clear that if he could stand, even barely, he would. "You say that'eh they have'eh money to put'eh me in'eh, in'eh some'eh place'eh far away but'eh –"

"They are not so –"

"Let'eh me SPEAK'eh! You tell me that'eh I need'eh to go so far. For this'eh there is'eh money? But'eh, to help'eh me to stay here there's'eh not'eh money? What kind'eh thing is'eh that'eh?"

"I am sorry, Nee ko lo." The care Martin takes to control his pronunciation is condescending. "What can I tell you, if you won't let me speak?"

Scattered rustling reveals the audience's discomfort. No one appreciates the antipathy of the dialogue, yet all recognize its relevance. A feeling of ill ease has pervaded the room. Its target is immaterial. This is not the resigned acceptance that Martin had hoped to instill.

"You talk about'eh where we must'eh go. What'eh someone else'eh say we must'eh do. That's'eh not'eh RIGHT'eh!" Barbarelli exclaims. His hard won suppression of familial dialect has been undone, uncovered like a false blonde's true roots.

Martin stands silent, hands loose at his sides. He would prefer the bored, superficial attentiveness of undergraduates, it seems. Barnard is momentarily sympathetic.

"This is a good point to stop," Martin finally states. "We can all look forward to Doctor Cordner's commentary at the next session." He lowers his head and proceeds without pause. "Thank you, oh Lord, for this meeting and for the understanding You have afforded us." He has found the abbreviated, irrevocable conclusion to today's Current Affairs that he needs.

Barnard remains seated amidst the general stirring. He looks at Phil expectantly.

"Are you going home, Phil?" he asks when they rise and walk out.

"Yes, I guess so. It's not that late. I can still get a few things wrapped up."

"You have some reading to do, then?"

"Yeah. Some. There are VieGie things I need to check on. I'll see you tomorrow. Want to meet for an early breakfast?"

"Uhhh, no. Not tomorrow," Barnard decides. "I want to sleep in. Haven't had a good night's sleep in a while."

Phil fist-bumps Barnard on the shoulder as they approach the elevator.

"Well. Hope you do tonight. Don't worry too much about your upcoming lecture. You've probably got it all worked out already, since you've been itching for the chance."

Barnard smiles and watches his friend proceed toward Sunset Vistas' exit. He turns away then taps the up button with his knuckle. He remains in place when the elevator door opens. Phil is only partially correct. Snatches of what he should and will convey tumble about in disorder. He stares up into the blue gloom of the empty box as its door slowly closes, anticipating another sleepless night, this time potentially more productive. His task is to organize a host of details that could obscure as much as enlighten. Therefore, he must create a network of relevant, interrelated facts, a web of rationale spun out upon the bridge strands – the mainstays – of circumstance. If he does so in bed, any perceived clarity will be the tantalizing phantom that succumbs to morning's light. Not the best way to proceed, he considers. He should think of the "assignment" as an opportunity and give it additional thought while the topic is still fresh in his mind.

Still early. Get some air. Might help getting to sleep.

He should start on it this evening and not expect it to simply come to him. He will review his note files, spend an evening or two downloading data and organizing, perhaps even go to the SMC library for a few hours or to a quiet bench at the beach front, and so set it down. After several moments of indecision, he walks to the entry for a brief solitary stroll in the damp coastal air. It may help him sleep through the night.

The truth, he recognizes, is that the effort will be more for his benefit than for theirs. Reason has only superficial relevance when need is current and acute.

END OF VOLUME 1

Made in the USA
Lexington, KY
11 November 2019